I0224341

SOUL *of the* KINGDOM

Biological Process and the Structure of Consciousness

SOUL *of the* KINGDOM

Biological Process

AND THE

Structure of Consciousness

Samuel Avery

Hart County, Kentucky

© 2019 by Samuel Avery
All Rights Reserved
ISBN: 978-0-9741976-2-3

I DEDICATE THIS BOOK TO MY PARENTS:

To my mother for teaching me to work,
To my father for teaching me to think,
To both for teaching me science.

CONTENTS

I   THE FORCE OF LIFE   1

II CELLS   13
    The Box, 13
    The Soup, 19
    Prokes and Eukes, 25
    The Cambrian Explosion, 38

III PHYLA   51
    Sponges, 51
    Hydras, 61
    Planaria, 65
    Relativity and the Cuttlefish, 70
    Ants, 77

IV THE STRUCTURE OF CONSCIOUSNESS   103

V   PLUMBING, FORTIFICATION, TELEGRAPHY,
    AND REPRODUCTION   119
    Plumbing, 119
    Fortification, 144

Telegraphy, 160
Reproduction, 191

VI HIGHER REALMS 221
The Observational Realm, 223
The Practical Realm, 232
The Equivalence of Self, 236
The Screen, 238

VII DEATH ON EARTH 247
The P-K Event, 248
The K-T Event, 260
The Modern Event, 267
Scenario One: Stasis, 281
Scenario Two: Catastrophe, 283
Scenario Three: Dominion, 284

VIII THE KINGDOM 291
Structure, 291
Reality, 298
Soul, 307

Bibliography, 313
Endnotes, 319

I

The Force of Life

When my children were small, I told them I knew magic. They gathered around – Jacob, Lauren, and Ben – wide-eyed, expectant, waiting to see what Daddy would do. Magic was a fun thing: exciting and a little scary. It was different from real, they knew, but there was no clear separation between magic and real. Their little half grins told me they were not about to put up with any tricks, but they wanted to see what daddy would do.

It was a spring morning and I was working in the garden. The kids were playing in the yard nearby when I called them over. I opened my fist, showing them a handful of bean seeds, and asked them to count. "Ten," said Jake, the oldest. Lauren, taking a few moments for independent verification, agreed. Ben, the youngest, smiled and slowly counted with his finger. "Ten!" he giggled, before finishing. "Now," I said, closing my hand, "I'm going to turn these ten beans into a hundred!" Their faces brightened. A hundred is an unimaginably big number, maybe more than you could count. "Follow me over here!" I beckoned. They trailed behind, still believing, but Jake already suspecting a trick. "Now, this is going to take a while," I admitted, "But these ten beans are going to turn into a hundred! Watch!" I placed all ten carefully in freshly turned soil, one by one. "Oh, dad!" Jacob said disappointedly. I covered the seeds and tamped the earth lightly. "You're just planting them," Lauren observed. The two oldest ran off, leaving only Ben to help me water. "It's magic!" I told him. "These beans are going to become a hundred! All we have to do is wait until the end of the summer." He smiled quizzically and ran off.

The rains came that summer, the seeds sprouted and grew, the sun shone, and by August the seedpods were dry and ready to pick. I coaxed the kids back to the garden. "Remember the ten seeds? Let's count them now." I picked the pods and shelled them into a bowl. "Now, let's count." "But, Dad," Lauren insisted, you just planted them and they grew." Her look was of expectation, as if I still might do something extraordinary. "It's still magic," I pleaded… "sort of." Ben looked into the bowl and stirred the beans with his finger. "It's not really magic," Jake observed. "Because they're not the same beans." "Oou…" I sighed, caught in the act. "The rain made them grow," he added.

"It's sort of magic, Lauren conceded, but you didn't really do it." "OK, I tricked you," I smiled. "The sun and the rain did it." They giggled as if at a bad joke. "Is it real magic?" Ben asked. "It's a different kind of magic," I explained, "but it's still magic."

Life *is* magic. We do not know where it comes from or how it works. That makes it magic. Like other forms of magic, life has properties beyond the limits of physical reality, but unlike other magic, it does not violate the laws of physical reality. There is an essential quality to life that non-life does not have, yet life occupies the same world. Living bodies bounce off each other, convert energy to motion, and fall to the earth the same way nonliving bodies do. They have mass, pass through time, and occupy space. They grow, move, reproduce, and respond to their surroundings within the laws of physics, but do so in ways that would be magical if accomplished by nonliving objects. We may come to understand the physical part through the scientific method. The other part of life, its non-physical dimension, we will never understand through the scientific method. Science does not and will not go there, no matter how hard we push. This part of life we will understand only through being alive. There is only seeing it, touching it, and being it. Where we would have understanding, we should have wonder. Wonder is, I believe, a most profound form of understanding.

Every spring, in the temperate zone where I live, I see the force of life pushing up through the Earth.

The force of life is a flowing stream that spills and shatters at the edge of oblivion, re-gathers, turns back, to the side, up, and over, becoming what it never was. It transforms bulk and magnitude into new dimension, meets obstacles, destroys what it is, dies, or evolves. It is loose proteins breaking, dissolving, churning against each other until, from nothing, a perimeter forms, cells bouncing together trillions of times until one day they stick; people streaming to towns and cities, crossing continents and oceans, dying by the thousands, creating society. It is algae marooned on dry land, seeds sprouting in soil, dinosaurs turning into birds, reproduction, thought, hunger, work, civilization, and the Big Bang: a linear progression colliding with the inevitable, becoming volume and shape: order becoming itself. It is a stream that winds and wears its way through time and space to higher form and meaning, flowing through microbes, insects, trees, and us, as we forge an everyday living from sweat and ingenuity.

In the outer world the force of life is seen in biological process and function, in organic evolution, plant growth, and animal behavior. Every day it grows, expands, feeds on itself, dies, and returns. In time it generates new species and families, new classes and phyla. Rarely, it generates a new kingdom. Within the human mind, the force of life is seen in planning, thinking, and working; it is felt in hunger, love, and anxiety. In this book I look at the force of life in both the outer world and within the mind, but I am particularly interested in the mind. I try to understand life from the standpoint of conscious experience, and proceed from that point to understand biology. We think of consciousness *inside* a physical body, subject to physiological process and related to biology through cause and effect. We tend to think that an object streaking across our line of vision or bouncing off our skin precipitates a chain of neurophysiological events that leads to the experience of seeing and feeling. We think of experience originating in the objective world. But I look at it differently. I try to understand seeing and feeling as existing in their own right – without causality – and I think of the objective world, the world of space and time, as arising from experience. Life comes first, then physics. Life does not evolve from the objective world; the objective world evolves from life. More specifically, I think the structural relation between seeing and feeling is what *creates* space and time. Consciousness is not in the dimensions; the dimensions are in consciousness.

Consciousness is a wholeness that comes in interrelated parts. The parts are separate *realms* of consciousness: both perceptual and conceptual. Seeing is distinct from thinking and hearing, and smelling from touch and taste, but each exists only as part of the whole. Each perceptual realm consists of distinct forms of information. There is no confusing one with another. Seeing is patterns of photons, hearing of vibrating air molecules, touch of electron repulsion, taste of direct chemical sensation, and smell of airborne or waterbourn chemical sensation. Aha! you may think. You have already caught me going outside of consciousness, speaking of photons, chemicals, and air molecules bouncing about in an independently existing world. But no, I speak of these *as they are experienced*. A pattern of photons is *what seeing is*, as

I mean it here. There are no photons without seeing: Light is visual consciousness. Air molecules are not little balls bouncing around empty rooms, they are *hearing*. They are bits of information of which the experience of hearing consists. Sound is auditory consciousness. Air molecules may be bouncing around a room – even a room with no one in it – but the room will exist only because there is hearing or potential hearing. Air molecules are little pieces of the *auditory realm of consciousness*. Singly they are nothing; it is in the order of their wholeness that perceptual consciousness exists.

Life evolves mostly as improvements in biological process: new and better methods for metabolism, mobility, defense, and reproduction. Organisms become larger, more complex, and more specialized. Occasionally, life goes further; it leaps to higher levels of consciousness by developing new realms of perceptual experience. This happened 545 million years ago with the evolution of the animal kingdom. Eukaryotic cells learned to function collectively and to *experience* collectively. Cells, as individuals, were limited to the chemical and tactile world of their immediate environment. As they became integrated cell communities and then animals, they found themselves living in a much larger dimensional world. The larger world was not experienced by single cells; it was sensed by the animal *as a whole* through the mediated experience of sensory cells. Retinal, cochlear, and olfactory cells relayed their experience to the rest of the animal body through encoded sensory information. Each type of information was transmitted to other cells through a separate medium, and each medium became a distinct realm of perception: seeing, hearing, or smelling. As experience became more collectivized within a cell community, animal consciousness evolved into organic wholeness.

Something similar is happening now among human beings. Much of what we experience as individuals is mediated through the experience of others. We know what is happening outside the house, or in another room, as much through what someone else *tells us* of their seeing and hearing as through our own direct seeing and hearing. We listen to what they are saying and see through their eyes. On a larger scale, electronic media have made us aware of what is happening in other places around the world. Our own eyes do not have to see London Bridge, sub-Saharan Africa, or glaciers melting in the Arctic; someone else's eyes and cameras have already seen them, and packaged what they saw for our use. Our bodies do not have to go anywhere to see the world.

This is an entirely new structure of consciousness. It began with the evolution of language sometime in the Pleistocene era and proceeds at an accelerating pace through the electronic revolution of the modern era. It is not a simple extension of the animal kingdom. It does not add a sixth sensory realm. It is, instead, a new dynamic of organic sentience among human beings built on language and video imagery. Through the possibility of all people seeing indirectly what any one person sees directly, it creates a higher order of consciousness. I call this the *observational realm of consciousness.*[1] Observation is not the same as perception – it is not what you or I see – it is what any observer *potentially* sees. Observation is potential perception.

This is not a scientific theory. A large and growing body of physical and biological evidence points directly toward it, but there is no evidence that proves it absolutely, and there never will be. Scientific proof is itself observation. It cannot see and test observational consciousness because that is what it is.

Science limits itself to the observational realm. It includes only that which can be repeated and verified by anyone at any time under the same conditions. It does not include what you think or what I see directly. It systematically excludes conceptual and direct perceptual experience. Science is not, therefore, everything: It is one realm among many. Because it excludes other than objective experience, it cannot see the larger picture of which it is a part. The view presented here is, therefore, by definition outside the bounds of science. It includes the dimensional world but is not of it. It is a view of life that includes all realms of consciousness, and a view that includes the viewpoint.

Though I emphasize the role of consciousness in this book, the next two chapters primarily concern biological process. Chapter Two describes the formation of cells from complex chemicals and the formation of multicellular organisms from cells. The first kingdom[2] of life is of simple cells without nuclei, known as *Prokaryotes*. These exist today as bacteria and certain forms of algae. The second kingdom, the *Eukaryotes*, consists of more complex cells with *organelles*: nuclei, mitochondria, chloroplasts, and other specialized bodies that perform functions of reproduction, metabolism, photosynthesis, etc. within the cell. It has been shown recently that at least some organelles have evolved from prokaryotic cells that were subsumed symbiotically by eukaryotes. Mitochondria, formerly free-ranging prokaryotes that became organelles hundreds of millions of years ago, still retain their own bacterial DNA. Reductively speaking, eukaryotic cells are integrated prokaryotic communities with a common outer membrane. They include Amoebas, Paramecia, and Euglenas, and every cell of the plant, animal, and fungal kingdoms. All "higher," multicellular forms of life consist of eukaryotes. During the so-called *Cambrian Explosion* of 545 million years ago, complex colonies of eukaryotic cells became the animal kingdom.

Chapter Three is about *Phyla*, the basic body forms assumed by cells as they come together to form animals. The discussion is not meant to be comprehensive. I include only the most well known of thirty-some animal phyla generally accepted by the modern biological profession and explore only a species or two in each phylum. I mention the plant and fungi kingdoms only in passing. My purpose is to show that as proteins become cells, cells organisms, and organisms societies, the force of life seeks higher levels of order not only within living beings, but between them. Life, though it appears confined to the space *within* living organisms, manifests itself also in the aggregate behavior of organisms. This has particular importance in understanding higher levels of human organization.

Chapter Four tackles the very difficult problem of the structure of consciousness. At what point does the "being" of individual cells become the "being" of an organism? Is a sponge an organism or a colony of cells? Is an ant colony a group of individuals or a superorganism? One tends to think of consciousness as a mysterious substance *within* a cell or an organism, but I will try to show that it is much more. I suggest that consciousness is all that there is, and that self, life, and the physical world are manifestations of a certain distinct structure of consciousness, a structure that is partly revealed in the relation between cell and organism, organism and society. I will contend that everything is understandable in terms of conscious experience, whether direct personal experience, or indirect experience available through science. For many readers this chapter will require the most patience. We are not used to analyzing consciousness – dividing a fundamentally mysterious phenomenon into separate realms and discussing their interactions – but I try to do it. I believe that the force of life is moving now as much in the conceptual as the dimensional world. The evolutionary process is taking place in the mind as much as in physiological process. As we watch the current ecological crisis unfold in the world around us, we see the physical basis of life diminishing, even as human population continues to grow. We sense the coming disaster. But this is only the physical basis of life; we are beginning to see the mental and spiritual bases of life as well. The discussion on consciousness resumes in Chapter Six.

Chapter Five describes an organic understanding of the evolution of human technology and its relation to human social evolution. There is a parallel between the evolution of organisms from cells and of societies from organisms, but the relation is imperfect and incomplete. It is an analogy, not an allegory. There are distinct similarities between nervous systems and electronics, between blood circulation and automobile traffic, between excretory systems and municipal plumbing, but there is no one-to-one correlation.

In Chapter Seven, I discuss long periods of ecological evolution and the role of mass extinctions, particularly of the P-T or *End Permian* extinction of 250 million years ago and the P-K or *End Cretaceous* extinction of 65 million years ago. Ecological catastrophes of the past give us deep perspective on the human-caused extinction event of the current era. I suggest three scenarios for the future of humanity and of the biosphere as a whole.

In the last chapter I try to make some sense of what life is doing now and where it may go in the future. The force of life, with its human manifestation in the forefront, is attempting unprecedented levels of new complexity. This has happened before, notably at the formation of the first fully enclosed cell membranes and again at the time of the Cambrian Explosion. What is different now is that the entire biosphere is caught up in the gamble. Where life has before gambled with individual species and whole phyla (losing many times before finding a winner), it now seems to be placing most of its chips on a single throw of the dice. Humans are only one of the chips, but it is humanity that has been entrusted with the dice.

Nothing like this has happened before.

II

———

Cells

THE BOX

Let us first look at life as it appears in the dimensions: within space and time.

Picture a box so enormous that planets and moons and stars whirl about inside it. Picture the box first devoid of life, with rocks, particles, and galaxies of all sizes streaming through space, colliding, accumulating, melting together, and disintegrating. Asteroids fall into planets, dust clouds gather into stars, atoms bond into molecules. Over time, molecules break apart, boulders roll down hills, fires burn out, stars explode, high concentrations mix with low concentrations, and things generally wind down. The orderly energy of large objects in motion gives way to the random motion of smaller particles. Energy remains constant, even as kinetic energy becomes heat. In the box you are picturing there are places where things are not going downhill as fast as everywhere else, but there are no places where things are going uphill.

Now, picture the same box with life. Order continues to decrease in the box as a whole, but it increases in a few special locations. Here, things wind up. Cells divide, seeds sprout, grass grows, children get taller, and the gross domestic product is up for a second year in a row. Overall energy remains constant, as before, but in living systems the disorganized energy of heat becomes the deliberate motion of living organisms. Living things seem to get bigger, better, and more complex. The difference between the box before and after life is a local increase in order over time. *Life is order.*

We measure living objects in the box the way we measure non-living objects: in increments of space, time, and mass. The box itself is constructed of space and time. Mass, the measure of what we call "matter," is manifest in the box as resistance to acceleration, which we measure in meters and two dimensions of seconds. (Mass is also manifest as a *curvature* of space-time, and at dimensional extremes, as a condensation of huge amounts of energy, but that is hard to picture in everyday life.) All of the physical yardsticks we use to make sense of what we see in the box – size, shape, velocity, temperature, pressure, acceleration, weight, momentum, wavelength, energy, force, etc. – are expressed in terms of space, time, and mass. All of the interactions between atoms and particles, planets and stars, tables and chairs, cells and molecules, are expressed in the same terms. What we cannot measure in the box is order. We cannot put a number on it. A living oak tree looks more orderly than a pile of mixed leaf, root, and cambium cells, but in purely physical terms, that is what it is. A living cell behaves in a more orderly fashion than a clump of carbon, hydrogen, oxygen, and nitrogen atoms, but there is no way to measure behavior in absolute physical terms. The best we can do is say that order is *improbable* – that it goes against the universal trend toward increased randomness – which is another way of saying that it is *not disorder.*

Whether or not we can put a number on order, we know it when we see it. A deer running through a forest, an amoeba wriggling under a microscope, or a child reaching for its mother all look orderly and alive. Something in the way we perceive them tells us that what is going on is more than ordinary physics. Whether or not we can say what order is, there are some things we can say about it. Most obviously, it is not everywhere. Much of the physical world has no order at all. Even a living system like a forest has elements that are not alive and that have no order in themselves. Objects and systems that have order can lose it. Things die. Even if they do not die they often lose order over time. A grizzly bear grows thin and weak with hunger, a forest turns dry and leafless in the autumn, or people stop mowing their lawns and a neighborhood becomes depressed and rundown. Life loses order at times, though its overall tendency is to increase it. Finally, and less obviously, order is not orderly to everybody. It comes only with a point of view. Order to one form of life may be disorder to another. From the standpoint of a vulture circling overhead, order would be the grizzly bear not finding food. From the standpoint of the grass it is more orderly when people in the neighborhood do not mow their lawns. There is no universal perspective from which to see order, at least in the box. There is no single *focus of order*; there are instead conflicting and overlapping focal points. We see order in mowed lawns and gross domestic products because we are human, and in healthy deer and grizzly bears because we are mammals and not vultures or intestinal parasites. Order is as much where you are looking from as what you are looking at. Because order is an experience, and not physical, it does not entirely fit in the box. When shoe-horned into space-time, this larger shape of living things becomes dimensionally flattened.

But the question we are after is, *How did life get in the box?* How and when and under what conditions did order arise from the random collisions of lifeless physical objects? One answer is that somebody put it there. Before closing the lid whoever built the box added life at the last minute. Another answer is that life arose on its own within the box. Things just naturally got more and more complex and life is that complexity. Both answers make good sense, the second requiring no thinking beyond the box. Another possibility is that life *did not* get in the box. The box is, instead, in life. The box is a dimensional framework that life projects into the world to understand itself and other physical phenomena. I suggest this as a means of compromising neither God nor science. Of these, the belief in science remains undiminished by elaboration.

Conscious experience is only partly of the box. Seeing, hearing, touching, etc. are of objects in space and time, but there are other forms of experience that are non-dimensional. Dreams, thoughts, prayers, imaginings, feelings, and emotions do not take place in the box. They may be shoehorned into the dimensions by turning them into elaborate electron dances in the cerebral cortex, but I do not think there is an honest way to make them into objective phenomena. It is simpler to appreciate them directly as they appear and disappear in conscious experience. Even the dimensional portion of consciousness – perceptual experience – is of the box but not in the box. Perceived objects are clearly in space and time, but the experience of perceiving them is not. Again, it is possible to squeeze consciousness into the box by reducing it to the interplay of neural impulses, but this merely removes the dimension of life altogether. Science gets into trouble when it goes beyond the limits it has set for itself, when it tries to put life inside of space and time. Life does make its appearance in the box, but only in dimensionally diminished form. That is what we see as *order*.

If this is true, I should revise an earlier definition. I said before that *Life is order*. It would be better to say: *Order is the way life appears in the box*. Life is not what is in the box; it is *looking* at what is in the box. It is experience.

In the following chapters we will look at life both inside and outside of the box. Inside we will see life creating new order by learning from its mistakes. Like non-life, it is subject to the disintegrating effects of random events, but unlike non-life, when it runs into trouble it makes copies of itself and tries again at other times and places. It makes orderly selections from random events in what looks like a mechanical process and weathers the storms of natural disaster, coming back bigger and better than before. It comes on again and again in a powerful stream of proteins, cells, organisms and societies, creating new levels from older building blocks, building on top of what it has already done. Life makes things happen. It does things. Non-life can only have things done to it.

From outside of the box we will see life in its entirety, box and all. The view is broader from here. But as life is beyond the organizational structure of space-time, the view will not be as clear. The picture will be less familiar and more difficult to discern. Without the discrete dimensional clarity of physical objects, and the certainty of place and sequence, it will be more difficult to say and understand what is happening. We will try to find structure even here, and suggest how that structure makes better sense of what we see going on in the box. We will look at what *doing* is from the standpoint of thought. We will say that the focal points of order we see in the box are just that, and not separate consciousnesses. We will try to distinguish between consciousness and *self*.

We begin in the box with how lifeless molecules interact to become living cells.

The Soup: Monomers and Polymers

There is nothing earthly about the building blocks of life. They are found on other planets, on moons, comets, and asteroids, and are present in great quantities throughout interstellar space. We find them today on meteorites that hit the Earth after swinging through the solar system for billions of years. They were floating in Earth's oceans before life began. The building blocks of life are *monomers*, or small organic molecules like amino acids, sugars, and nucleic acid bases, themselves made of CHON, or Carbon, Hydrogen, Oxygen, and Nitrogen atoms. They are the bricks of life. But they are not the building. They are not alive by themselves, and are easily synthesized in the laboratory. What brings them to life is the manner in which they are arranged. Life is the order in which the bricks are placed. DNA chains, bacteria, jellyfish, sequoias, and elephants are monomers piled together in interesting ways.

Monomers are made of Carbon, Hydrogen, Oxygen, and Nitrogen, rather than, say, of Iron, Vanadium, Boron, or Iridium because CHONs bond easily to one another and are four of the five most common elements in the universe. (The fifth is Helium, which does not bond at all and is therefore not found in monomers.) Monomers are seasoned with a sprinklings of phosphorus, sulfur, calcium, and other atoms, but remain 99.9% CHON. Examples of monomers are a simple sugar, glucose ($C_6H_{12}O_6$: 6 carbon, 12 hydrogen, and 6 oxygen atoms), glycine (H_2N-CH_2-COOH), an amino acid, and acetaldehyde (CH_3CHO), a component of DNA. Life begins to take shape when monomers are linked together, one at a time, into long chains, or *polymers*. Polymers are buildings made of monomers, and *polymerization* is the brick-by-brick construction of biological process.

Three prime examples of polymerization are glucose monomers linked into *carbohydrates*, amino acids linked into *proteins*, and nucleotides linked into *nucleic acids* like RNA or DNA. Each process is distinctly orderly, as if in fulfillment of a plan, yet each is explainable in terms of cause and effect and there is no *visible* agency of intelligence. Carbohydrates are the fiber of the plant kingdom and the food of most living things, proteins are the meat and muscle of all living things, and nucleic acids are the architectural blue prints that determine what type of living thing monomers are built into. But polymers by themselves do not constitute life because they easily break back down into monomers. Long, elegant polymers may by chance coalesce from randomly interacting monomers, but left unprotected in the primordial soup, they do not stay elegant for long. Shoved and torn by other molecules in the soup, they break apart too quickly to do anything life-like. They need a barrier isolating them from the chaos of the outside. Even this is not enough. To keep making new order they need a constant source of energy. Then, to become life, they need to find a way of replicating the order they have achieved, a way of passing on what they have become to some other time and place. To become life, polymers have to be able to save the hard work (or the good luck) of achieving orderly status by developing an outer membrane, a metabolism, and reproduction. They must create, maintain and increase order through time.

Organic monomers like amino acids, glucose, and even some nucleotides were in the box before the advent of life. Polymerization of monomers into complex organic molecules, though extremely rare, might also have happened at places in the box before life. Cell membranes probably happened a short time after the beginning of life. Life began some time between polymers and membranes, somewhere in the primordial soup when large organic molecules clumped together and protective barriers began to enclose them. "Naked genes" composed of primitive Ribonucleic acid (RNA) were already beginning to replicate portions of themselves. They could not make complete copies of themselves, like the Deoxyribonucleic acids that came later (DNA), but were able to "copy and paste" portions of gene sequences that directed future polymerization of monomers. Clumps of complex organic compounds were able thereby to generate new clumps that resembled them. Before there were living cells these naked genes were what might be called "living molecules."

Large organic molecules are insoluble in seawater. Like oil droplets in water, they do not distribute themselves evenly throughout the medium, but cling together in small formations. Some molecules are polarized, like molecules of soap, with a *hydrophilic*, or water-loving tendency on one end and a *hydrophobic*, or water-fearing tendency on the other end. As they clustered together in open seawater these large organic polarized molecules arranged themselves with their hydrophilic ends outward and hydrophobic ends inward, forming spherical bubbles (again like soap). Later, a second layer of molecules developed head to head with the first, their hydrophobic ends linking with those of the first layer and their hydrophilic ends pointed to the interior of the bubble. Packed together tightly, they formed a watertight film, enclosing other larger, more complex organic molecules at the center of the bubble. Proteins (custom designed polymers built from amino acid monomers) were later embedded in the film to stiffen and strengthen the membrane. Other proteins provided pores in the membrane that allowed nutrients in and waste products out. All living cells have this basic membrane structure.

Membranes became more sophisticated as the nucleic acids they protected synthesized phospholipids, or molecules with a "head" of carbon and phosphate at one end and two long "tails" of fatty acids on the other end. Tails had to be made to just the right length: too short, and the membrane leaks, too long, and nothing passes through. Fatty acid tails were designed, presumably through trial and error, to be long enough to keep "their own" larger organic molecules in, and "foreign" molecules out, while allowing smaller metabolic molecules to pass in and waste products to pass out. This became an extremely demanding task as more complicated cells evolved with more complicated nutritional needs. Membranes evolved originally from natural compounds, providing a simple layer of defense against chaotic forces outside the cells, but once subject to cellular control they became increasingly complex and sophisticated. By controlling traffic in and out of the cell, they developed a focus of living order within their perimeters.

Before cell membranes were well established, solutions of proto-cells with primitive leaky membranes allowed proteins, carbohydrates, and nucleic acids synthesized in one cell to become available to other cells in the primordial soup. This no doubt stimulated evolutionary creativity at first, as the right protein from another proto-cell might come floating by at just the right time. Eventually, more effective membranes put an end to this, as cells with the best DNA, the best metabolic system, and the best enclosure chose to keep what was theirs, to survive when other cells dissolved, and to reproduce themselves exactly as they were. In fact, whatever the variety of early cells and proto-cells, all of life still living originated, as best we know, from a single successful cell. Some super special combination of semi-permeable membrane, polymer synthesis, and metabolism replicated itself so effectively so many times that it became the ancestor of all later cells, and thereby of all life on Earth.

How did one cell and its offspring out-compete all other cells and proto-cells? Most likely, it ate them. It did not bother killing them; it just waited until they fell apart. Then it assimilated their monomers into its own daughter cells. Living organisms still digest food by breaking down proteins and carbohydrates (polymers) into amino acids and glucose (monomers) before sticking them back together their own way. They break their food into bricks, then use the bricks to build more of themselves.

It was originally thought that the first living cells had to be *photosynthetic*, that is, they had to make their own food by absorbing energy from the sun. Cells that do not make their own food could not come first, it was thought, as they had to have some other form of life to eat. But it is now understood that the first cells lived not by making their own food but by consuming monomers and other free-floating chemicals. As all types of cells and proto-cells formed and disintegrated in the primordial soup, the first successful cells and their offspring consumed the glucose, amino acids, and nucleotides streaming past. They drank the soup. But this primitive form of chemosynthesis could not last indefinitely, as it depended on a non-renewable resource. As more and more cells formed, the supply of free-floating monomers dwindled, and cells were forced to develop other forms of energy production. To prevent starvation and extinction, every cell had to develop some form of chemosynthesis or photosynthesis. They had to find renewable sources of raw materials and energy. Solar energy was not a direct substitute for the chemical energy of compounds readily available in the soup; it was a long-range adaptation that required many steps to accomplish. Not all of life went solar, but the main trunk of life bent toward the sun.

The first few enclosed cells gradually consumed the environment that gave them birth. Their only way to live was to eat their way through a non-renewable resource base. To survive through the long term, they had to learn new tricks. But every cell that evolved and emerged from the primordial soup was a direct descendant of the first enclosed cell. All life on Earth remains, thereby, related to all other life on Earth.

PROKES AND EUKES

Life goes back a full 3.8 billion years, getting underway as soon as the earth cooled enough to allow it. The earth is about four and a half billion years old, and the universe as a whole, if you are keeping track, 13 billion years. To make sense of such long periods of time it is best, I think, to forget about minutes and hours and years – even centuries. It is nearly impossible to understand millions and billions of years by extending your own experience of the passage of time. If you try to do so, everything looks like "a long, long, time ago," or "an even longer long time ago," and you fail to understand the proportionate relation of say, a million years, in which not much happens geologically, to a hundred million years, in which a great deal happens. It is best when thinking in geological time to adopt a million-year yardstick by which to measure everything. There are 3800 of these yardsticks in the entire history of life on earth, 65 of them since the dinosaurs died, and 0.15 of them in the history of *Homo sapiens*.

For its first two billion years (2000 yardsticks) life consisted entirely of simple *prokaryotic* cells without nuclei, mitochondria, chloroplasts or other discrete structures with membranes inside the cell. Prokes exist today in the form of bacteria and blue-green algae. The fossil record is sketchy at this early date, but the first type of prokaryotic cells to evolve that still exist are the *thermophilitic*, or "heat loving" bacteria that adapted to the extremely hot and carbon rich climate of the times. 3.8 billion years ago the average temperature on the Earth's surface was above the boiling point of water. As the earth cooled and oxygen filled the atmosphere, thermophilitic bacteria scaled back their occupation of the earth's surface, but they live on today in hot springs, geysers, hydrothermal vents and other refuges that resemble environmental conditions at the time they evolved. In terms of total time served they are the most successful life form ever to exist.

Most prokes are chemosynthesizers. A few utilize a form of photosynthesis that does not involve oxygen. Most interesting from an evolutionary point of view are the few bacteria that learned to make a living by earning it themselves. They accomplished this with an oxygen-based form of photosynthesis. C*yanobacteria,* as they are called (otherwise known as blue-green algae), have made three fundamental contributions to the living world as we now experience it. First, they perform *all* oxygen-based photosynthesis on the planet. Second, by releasing as a waste product of photosynthesis billions of tons of oxygen into the air over hundreds of millions of years, they are responsible for creating the oxygen-rich atmosphere we breathe today. In providing an energy supply for themselves and for most other forms of life they simultaneously provided the metabolic basis for other cells to utilize that energy. Third, cyanobacteria removed most of the carbon dioxide from the ancient atmosphere, reversing the greenhouse effect and cooling the Earth's climate to a point where more diverse forms of life could develop. They eliminated carbon dioxide directly by absorbing it for photosynthesis, and indirectly by spreading out over the land, dissolving the surface minerals of rocks and releasing silicates into the environment. Carbon dioxide combined chemically with these silicates and was removed from the atmosphere. No other form of life has so drastically altered its environment. Only humanity has altered the environment to anything approaching these proportions. But the human impact on the atmosphere, though less extensive, has been many times more rapid. Where cyanobacteria changed the atmosphere gradually over two billion years, we have changed it in the last two hundred.

Prokes constitute the first life *kingdom*, the *Monerans*. (Simply stated, "Moneran" means "prokaryote," which means bacteria. Even "algae" can mean bacteria, in the case of blue-green algae. These are all names for single-celled organisms without nuclei.) A kingdom is the broadest category of living organisms. There are four kingdoms beside Monerans: *Protists*, or single-celled *Eukaryotic* organisms that I will discuss shortly, *Fungi, Plants,* and *Animals.* Each kingdom is further divided into phyla, or basic body types, and phyla are divided into classes. (I will discuss major animal phyla in the next chapter.) Biologists used to recognize only two kingdoms: plants and animals. This is based on what they saw, and on what you still see today: If you take a walk in the park you see trees, grasses, insects, worms, bushes, flowers, dogs, squirrels, etc., each of which is either a plant or an animal. Plants make their own food and are stuck in the ground where animals move around and get their food from plants or other animals. These distinctions become troublesome, however, on the microscopic level. Some animals, even multicellular animals, are rooted in the ground like plants, and some single-cellular "plants," like *Euglena*, move around but still photosynthesize their own food. Other organisms, like fungi, do not make their own food but look a lot like plants in other ways. Biologists decided in the later twentieth century to base the distinction between kingdoms not on how organisms appear on the macroscopic level, but on how they are structured. The fungus, plant, and animal kingdoms are all *multicellular.* They are distinguished from the single-celled kingdoms by their use of cells as building blocks to create higher levels of biological function and, as I hope to show, higher levels of conscious experience. All three kingdoms have developed organisms of such high cell specialization and interdependence that the organic whole has become more than the sum of its cellular parts. The animal kingdom is of greatest interest to the purpose of this book in that animals have developed sensory organs and what I will call *dimensional consciousness.*

Monerans (prokaryotes) and Protists (eukaryotes) are both unicellular. What structural difference could justify granting each an entire kingdom? A proke is a cell membrane enclosing a variety of polymers capable of, among other things, metabolism, protein synthesis, and self-replication. One of the polymers it produces from glucose monomers is cellulose, a strong but stiff fiber with which it coats its exterior surface, creating, in addition to the cell membrane, a *cell wall.* This gives more protection than the membrane alone provides, but greatly reduces flexibility. Prokes cannot change shape. They move, but cannot move very deliberately, and cannot engulf food particles. Internally, there are no specialized structures: no nucleus with a membrane separating genes from the rest of the cell and no *organelles* for digestion, metabolism, photosynthesis, or reproduction. Everything just floats around. (Prokes do, however, have ribosomes for protein synthesis.) Each biological function is managed by large, free-floating organic molecules swimming about the interior of the cell in more or less evenly distributed concentrations. Prokes are as simple as a form of life can be and still be life. They have mastered ways of linking monomers into carbohydrates, cell walls, proteins, and nucleic acids that have worked for them and kept them alive over the last 3.8 billion years. They have had little reason to evolve in any fundamental way. They are so satisfied with themselves that life would have evolved no further if it had to depend on prokes to come up with new ideas. Prokes don't want to do anything new. Evolution, rather than asking prokes to change into something else, would build new, larger structures using prokes as building blocks.

The new structures were not multicellular. Instead, they were a new type of cell. Prokes were swallowed whole and incorporated into the living process of an evolutionarily unique enclosed cell, the *eukaryote*. Eukes, like prokes, were single-celled, at first. The structural difference between them is quite simple: prokes live inside of eukes. Specifically, cyanobacteria (prokes) were engulfed by algae and plant cells (eukes) and became *chloroplasts*, tiny organelles within the larger host cell that carried on photosynthesis. Eukes are incapable of photosynthesizing on their own. Other forms of bacteria became other types of organelles. Mitochondria, formerly free-living bacteria, use oxygen to metabolize carbohydrates within nearly every plant, fungal, or animal cell on earth. Eukes are not, therefore, just bigger and better prokes; they are a separate form of life built on the backs of prokes. Allowing bacteria to live within their membranes, eukes have sheltered and nourished them in return for food production, metabolism, and other cell functions. The system is known as *endosymbiosis*, or a symbiotic relation in which one organism lives inside another. Endosymbiosis, like any symbiosis, can be interpreted as parasitism or slavery, depending on who is perceived to be using whom to get the better end of the deal. But because the arrangement has been stable for the last two billion years it would appear that both parties have benefited. The euke has concentrated on the size, shape, mobility, defense, and reproduction of the cell, and left energy production and metabolism to the proke. The system is so stable that each party can be said to have lost its separate identity. Cyanobacteria that became chloroplasts have given up more than 90% of their DNA to the euke's nucleus (retaining the remainder within the chloroplast). Mitochondria, thousands of which exist within most eukaryotic cells, have surrendered a considerably smaller portion.

Mitochondria reproduce themselves within the eukaryotic cell. When a multicellular organism reproduces itself it must carry not only its own cellular DNA, but mitochondrial DNA as well. Mitochodrial DNA is always transported to the next generation in the egg cells because it takes up too much room to be included in the much smaller sperm cell. Mitochondrial DNA has become a useful tool for tracing lineage, but it traces only female lineage.

Chloroplasts probably evolved in eukes at the same time as mitochondria because the oxygen they released within the cell was toxic to other organic compounds. Oxygen combines easily with other chemicals, releasing energy but destroying molecular structure. Mitochondria utilize oxygen in an orderly manner, releasing energy to the cell at a measured pace to keep the cell from burning itself up. Like a carburetor in an automobile engine, they deliver just the right amount of energy to the cylinder in just the right form without igniting the fuel tank. Eukes probably engulfed the bacteria that became chloroplasts and mitochondria some time in the evolutionary past much the same way they engulf food particles. Unlike prokes, eukes are able to change their shape by manipulating their cell membranes with an interior structure of fibrous material known as a cytoskeleton. Cytoskeleton filaments are attached to points on the cell membrane and contract within the cell to change its shape. As the membrane contacts a food morsel or a bacterium, it surrounds it on all sides, breaks off a portion of itself, turns it inside out, and forms a vacuole within the cell containing the morsel. No one knows the particulars, but at some point in their history eukes discovered that, rather than eating certain bacteria, they would be better off letting them live within the membrane. Evidence of the engulfing process remains to this day in the double membrane separating organelles from the rest of the cell's interior; one membrane originates with the host cell and the other with the free-living bacterium. As the cyanobacterium was swallowed it kept its own membrane even as it was surrounded by the host cell's membrane. Chloroplasts in *cryptophytic* alga indicate that the engulfing process happened on more than one level. With *four* membranes and the vestigial remains of a nucleus between the second and third membranes, these particular eukes must have engulfed other eukes – mitochondria and all – in a secondary endosymbiotic event. There is even a case of *tertiary* endosymbiosis among *dinoflagellates*, whose chloroplasts have *six* membranes.

The eukaryote's ability to manipulate its outer membrane may also prove important in understanding what it experiences. It is difficult to speak of what a bacterium or an alga cell "experiences," but it cannot be assumed that any biological process, however simple, is purely mechanical and without sentience of some sort. All life does things – all life creates order – and to create order there must be existence beyond the merely mechanical. No one can say what that existence may be and there is no way to know its content, but it is possible to know something of its structure. By structure I mean what realm of sensory consciousness a particular experience may fall into, how that realm is related to other realms, and how it may resemble realms that we, as humans, experience. Clearly, single cells can have no experience in the visual, auditory, or olfactory realms. But the biological processes they exhibit indicate that they do in some way experience the chemical and tactile realms. They do not "taste" the way humans taste, but they encounter chemical stimuli in their environments on a regular basis and have to distinguish between one chemical and another in order to know what to allow through the cell membrane and what to keep out. What we experience in our taste buds – and throughout our digestive tracts – is this same cellular absorption of chemicals from the surrounding milieu.

Cells experience chemical perception as molecules pass through the membrane. Some cells also perceive molecules, particles, and other cells that *do not* penetrate the cell membrane, but merely contact it on the exterior. This is the *tactile* realm of perception. Touch, to become an entirely separate realm of perception, depends on a more flexible and sophisticated membrane than that of the average bacterial cell. The proke, with its stiff, cellulose-encrusted cell wall likely has no idea what is going on outside. The euke membrane, on the other hand, with its far greater mobility and flexibility and its habit of surrounding and engulfing external objects, is in a much better position to make sense of tactile stimuli. It is reasonable to suppose, without knowing absolutely, that the rise of the Eukaryotic kingdom is also the rise of the tactile realm of consciousness. The structural revolution in biological processes that characterizes the evolution of eukaryotes from prokaryotes is also a revolution in the structure of perceptual consciousness.

It is generally assumed that eukes originated as simple cells, perhaps coexisting with prokes for millions of years and only later engulfing them. This is supported by the fact that there is a living euke, *Giardia*, that has no mitochondria, even today. *Giardia* could be a pre-endosymbiotic ancestor of other eukes. But, being a parasite, it is equally possible that it had mitochondria in its evolutionary past but gave them up once it inhabited host organisms that made them unnecessary. This brings up the possibility that eukes evolved from the very beginning *only* as combinations of prokes, that is, that they never had any sort of separate existence. Recent research has shown that *Girardia* and related cells all carry bacterial genes in their nuclei, indicating that all known eukes participated in some form of endosymbiosis with bacteria in the past, whether or not they do so now.[3] It is possible that all parts of the euke cell, membrane, nucleus, mitochondria, choloroplasts, ribosomes, golgi apparatus, etc. are all of prokaryotic origin, and the eukarote itself is no more than the whole over and above its bacterial parts.

Eukes first evolved over 2 billion years ago, but did not proliferate in a big way until 800 million years later. Throughout this long "fuse period" prokaryotic bacteria continued to rule the waves, and even the rocks. The innate simplicity of prokes gives them great advantages. For one, they are much more metabolically diverse. Where eukes are almost entirely dependent on oxygen-based photosynthesis and respiration (with some using anaerobic fermentation), prokes retain a wide variety of anaerobic (oxygen-free) and chemosynthetic forms of metabolism. Some of them can even change from aerobic to anaerobic, depending on conditions. Less specialized, they live longer as species. Some fossilized bacteria species known to live billions of years ago are alive today. Proke species seem to last almost indefinitely, where single-celled euke species last for a mere 100 yardsticks or so before going extinct. Specialization pays an even higher price with higher, multicellular eukes. (All plant, fungi, and animal cells are eukaryotic.) The average life span of animal species is only about four million years. Proke cells reproduce by dividing autonomously, without regard to their neighbors, do not become interdependent, and have not evolved into multicellular organisms. Eukes, on the other hand, use molecular signals passed from cell to cell to choreograph cell reproduction into complex patterns of division and differentiation. This ability makes multicellular growth possible (95% of euke species are multicellular.) But intercellular dependence increases vulnerability to extinction. Specialization and higher levels of organization mean more rapid evolution, but they also mean greater risk of exposure to changes in climate and other environmental conditions.

Conditions favorable for a eukaryotic takeover of the Earth had to wait until cyanobacteria had released enough oxygen into the atmosphere to support aerobic respiration on a large scale. Oxygen is also necessary for the absorption of nitrogen, a necessary nutrient for protean formation. The Earth's atmosphere is 78% nitrogen, but plants cannot absorb it directly. Cyanobacteria *can* absorb it directly, but those that evolved into chloroplasts within eukaryotic plants cells were no longer in direct contact with the atmosphere and could not perform this service for their hosts. Plants instead have to absorb nitrogen when it combines with oxygen, dissolves in water, and becomes "fixed" in the soil by free-living bacteria. Prokes continued to rule the Earth for hundreds of millions of years while oxygen was building up in the atmosphere. Around 1.2 billion years ago there was enough oxgen to trigger the eukaryotic "Big Bang."

There are about 4000 named species of bacteria (though scientists generally name only those that are cultured in laboratories). There are close to two million euke species that we know of: 100,000 Protists (protozoa and alga), 100,000 fungi, 300,000 land plants, and well over a million animals, most of them insects. On the macroscopic level, prokaryotic bacteria seem to play a small role in the biosphere. But it should be kept in mind that all species of all higher kingdoms are entirely unnecessary to life. Eukes are but convolutions and ornamentation of the prokaryotic theme, built of self-contained prokaryotic modules that do not need to be built into anything. Prokes could get by just as easily without all the eukaryotic frills of organelles, multicellularity, and suburban subdivisions. As you look out the window at a flower, a lawn, or a forested hillside, the green you see is photosynthesizing bacteria, within cells, within multicellular organisms. Prokaryotes could be doing it on their own.

The Cambrian Explosion

From the Eukaryotic "Big Bang" of 1.2 billion years ago until the "Cambrian Explosion" of 545 million years ago, the force of life had over six hundred million years to think about what to do next. One possibility was to take what it already had, the eukaryotic cell, and make it larger. Another possibility was to take the same cell, leave it pretty much the way it is, but stick it together with other cells of the same kind in a long chain. Still another possibility was to stick cells together in two-dimensional mats and give them some metabolic advantage in communal living. Finally, the possibility arose of sticking cells together in such a way as to form their own enclosure. All of these possibilities were tried.

Eukes *without* chloroplasts became amoebas, paramecia, G*iardia*, and other single-celled "protozoa," and later evolved into fungi and animals. Eukes *with* chloroplasts became algae and later evolved on land into plants. But the first real attempt to do something interesting after the Big Bang was the appearance of an extremely large alga cell, the *Churia*. The Churia was a single-celled photosynthesizer up to one centimeter in diameter. You could see it and hold it in your hand. It first appeared about 1.1 billion years ago and peaked 200 million years later, but died out well before the Cambrian. It is not known how it got so big or exactly why it got no bigger, but it is one of only a very few examples of living organisms growing to macroscopic dimensions without some form of multi-cellular structure. It worked, but it did not go anywhere. The eukaryotic cell, rather than evolving by expanding its own dimensions, would become a building block for an entirely new form of expansion.

Some species of algae became multicellular millions of years before the appearance of animals, fungi, or land plants. But the type of multicellularity they developed consisted entirely of sticking on to one another. They stayed stuck, but metabolized, photosynthesized, and reproduced separately. There were advantages in this form of multicellularity or it would not have continued, but the degree of interdependence was minimal. It was not necessary for survival. Chains broke, individual cells floated off on their own, divided, and formed new chains. As long as each cell could produce its own food, and as long as each cell was surrounded by seawater, there was no need for specialization and no purpose in creating any sort of interdependent multicellularity.

Cells that did not produce their own food were in a different situation. There were distinct advantages in teaming up, even if each cell remained surrounded by seawater. About 700 million years ago, the eukaryotic world may have been on the verge of a multicellular breakthrough, but a series of glaciations occurred then that killed most of the living world. Another 150 million years would elapse before the multicellular revolution could begin.[4] This was not the Pleistocene Ice Ages (1.8 million to 12,000 years ago) witnessed by Stone Age humans – this was much worse. For some unknown reason, glaciers began creeping down from the poles into temperate and even tropical latitudes. Ice forming on mountain ranges spread down through river valleys and out across oceans. The more extensive the glaciers became, the more sunlight they reflected back into space, further cooling the Earth. This positive feedback loop further cooled the planet until nearly the entire surface of the Earth was covered with ice. We do not know how extensive the ice cover might have been. Perhaps life survived in a narrow ring of broken ice near the equator, or perhaps near volcanoes or around underwater hydrothermal vents. Photosynthesis came to a near standstill. Carbon dioxide trapped under the ice could not reach the atmosphere, and oxygen trapped in the atmosphere could not reach the ocean. Any complex multicellular life that might have evolved was surely destroyed. This Precambrian Ice Age ended abruptly, probably due to volcanoes spewing enough new carbon dioxide into the air to trap heat and re-warm the planet. Ice ages would return after the appearance of animals, but none was as severe, not even the one we experienced a few thousand years ago as Paleolithic hunters.

Over a hundred million years later (about 600 million years ago) a series of multicellular organisms appeared that were not like anything living now. These were "animals" in that they did not produce their own food, but they were mostly immobile and had no specialized organs or organ systems. *Ediacarans,* as they are called, came mostly in the form of large, disk-shaped, soft-bodied mats that lay draped across the sea floor. Many displayed cylindrical tube structures that are not fully explained. Like sponges and cnidarians, all cells were in direct contact with seawater for nutrition and waste removal, and there was no mouth, no head, no gut, no circulatory system, no digestive tract, no gills and no lungs. Each cell was pretty much on its own, much more so even than in the case of sponges. What was the advantage, one wonders, over the single life? It is not known how Ediacarans metabolized (that is, how or what they ate, but they probably carried on some form of symbiotic relationship with bacteria. Shallow water species provided a solid platform for photosynthesizing bacteria, gaining food indirectly from the sun, while deep-water species probably supported bacteria capable of metabolizing methane and hydrogen sulfide as it seeped up from the ocean floor. This would explain the large, flat Ediacaran body, and its advantage over single cellularity. These "animals," if they can be called by the term, existed for many millions of years in the predator-free seas before the Cambrian Explosion, but became extinct at that time or soon after.

With our innate human need to organize and categorize, it is difficult to know what to do with something like the Ediacarans. That they are multicellular is without doubt. Their degree of cellular interdependence is questionable, though some forms may have reached interdependence similar to animals. That they are built of self-contained living modules (eukaryotic cells) requires us to grant them, I believe, the status of kingdom. But what kingdom? Some paleontologists think they are early forms of the animals. A few species look like sponges, jellyfish, or worms, and others show mobility and bilateral symmetry. Some types even display what might be interpreted as legs, segments, cavities, calcareous skeletons, nerve nets, and muscles. But other paleontologists do not see this and see instead fundamental differences between Ediacarans and Cambrian animals. Either they were a separate branch of animals that became extinct before the Cambrian, or they were a separate kingdom altogether.

The discussion says as much about the human mind as about Ediacarans, but I think there is one important point to make. Unlike "true" animals of the Cambrian, Ediacarans, for the most part, did not form enclosures. They had no gut. They had no systematic defense from the outside. They had no mouths, teeth, stomaches, intestines, or anuses. There was no curvature to their body tissue that separated order within from chaos without. There was no internal or external, no sacred space inside or defensive barrier to the outside. Ediacarans, whatever evolutionary bridges they may have crossed, were pale reflections of what came next.

Some paleontologists believe that Ediacaran fauna were extinct 545 million years ago and that Cambrian fauna evolved in an empty world. Some think that Ediacarans survived but were quickly consumed by Cambrian predators. Others think that they survived and *became* Cambrian fauna. Nearly all agree that what happened at the beginning of the Cambrian period was truly explosive, both qualitatively and qualitatively. Hugh numbers of entirely new and diverse animal species appeared on the scene quite suddenly. (In geological terms, ten yardsticks is sudden.) For many scientists this is the beginning of real paleontology and geology: All animal phyla, living and extinct, date from the Cambrian and most of the Earth's exposed rocks are Cambrian or younger. So much more is known from this period forward than anything that happened in the four billion years before it (85% of the Earth's history) is often referred to simply as "Precambrian."

There is no general agreement on what triggered the explosion. Another minor ice age had just ended, perhaps clearing the decks of lingering Ediacarans. Algae were spreading and diversifying, creating new possibilities for grazers. The oxygen supply was up to nearly modern levels, and plenty of biospace was available in the Earth's oceans for experimenting with new ways of combining eukaryotic cells. Proteins excreted on the outer surface of cell membranes held cells together, creating body tissues, and membranes picked up chemical signals from other cells and relayed them to the nucleus. Cells learned to coordinate mobility, metabolism, and division into new cells, forming new tissues. Conditions were so ripe for creative experimentation that about 100 entirely new and different body plans were developed: some with heads and mouths, some without; some with digestive tracts, some without; some with legs, fins, tails, tentacles, gills, and body segments, some without; some even with eyes and primitive brains, and some without. Anything was worth a try. There was no good reason *not* to try a tentacle on the head or on a hind leg, a mouth in the middle of a digestive tract, or another layer of cells here or there. Not at first. Most of these first animals were scavengers, grazers, and filter feeders, extracting bacteria, algae, or bits of organic matter from the substrate or from passing seawater, but others became predators. Body plans had to take defense and offense into consideration. Prey species were forced to narrow possibilities down to those that did not expose soft body parts, and predators were forced to develop claws, suckers, stingers, drills, and teeth that could penetrate the new defenses. The result was a Cambrian arms race. Each side struggled to come up with new weapons and tactics to counter the weapons and tactics of the other side. The arms race meant new body forms and it meant new forms of animal behavior and interaction or, in other words, ecology. As available biospace became filled with life, each form of life had to evolve in relation to others.

A few examples of Cambrian species are *Olenoides, Hallucigenia, Opabinia, Anomalocaris,* and *Pikaia. Olenoides,* a trilobite, was several centimeters long with chitin exoskeleton reinforced with calcium carbonate on the upper side. Like all trilobites, it had three main body sections: a head shield, a series of body segments in the mid section (seven in the case of *Olenoides,*) and a posterior shield. Each mid-section segment had a pair of jointed walking legs and gills. It had eyes with silicate lenses and sensory appendages near the mouth that helped it find prey. Similar appendages in the posterior section notified the animal of attack from behind. With its exoskeleton and jointed appendages *Olenoides* was a member of the arthropod phylum.

Hallucigenia belonged to a separate phylum without re-lation, as far as we know, to any currently existing phylum. It had a one-inch elongated body with seven pairs of long sharp spines on one side and seven tentacles, or pairs of tentacles on the other. It was thought originally that *Hal-lucigenia* walked on the spines and caught prey with the tentacles, even though there was no distinct mouth and the tentacles were too far from what appeared to be the head. The possibility was suggested that each tentacle was its own separate feeding apparatus, and therefore one of seven mouths! But when a second row of tentacles was dis-covered scientists decided that the animal made more sense flipped over, with the tentacles for walking and the spines for defense. But problems remain: with what appears to be a large puffy sac on one end of the body and something like a nozzle on the other, nobody is sure which end is the head and which the tail. We are not surprised it was named *Hallucigenia.*

Opabinia also belongs to no known phylum, living or dead. It was about two inches long with 15 armor-covered body segments with gills on top and no legs. It was dis-tinguished from all other animals by a trunk-like nozzle extending from its head with grasping claws at the far end. Like an elephant, it bent this strange feeding organ to a backward facing mouth on the underside of the body. *Opa-binia* is further distinguished by having five eyes!

Anomalocaris, over a meter long and one of the larg-est Cambrian animals, was a predator and probably an arthropod. It had large eyes, body flaps for propulsion in water, and a tail fan for propulsion and steerage. Its walking legs were soft and flexible, a distinctly non-arthropod trait, but it had large jointed appendages near the mouth for grasping prey. The mouth was surrounded not by teeth but by rows of plates with sharp prongs that forced prey into the gut.

Pikaia was an elongate wormlike animal with a flattened body tapered at both ends. It had appendages near the mouth, gill slits, and a very small brain. Unlike other Cambrian fauna, it waved its entire body side to side as it swam, like a fish. To stiffen the body and help it spring back with each sideways motion, *Pikaia* had a long, tough, but flexible rod along its back, just within the body. This organ is a *notochord,* and clearly identifies *Pikaia* as a member of the phylum Chordata, and as the ancestor of all vertebrates: fish, amphibians, reptiles, birds, and mammals.

The primary resources for both predators and prey in the Cambrian arms race were dissolved minerals. Calcium, magnesium, and other minerals were extracted from seawater, combined into carbonates and silicates, and excreted to form a hard protective armor around an animal's body or parts of the body. Some was shaped into shells, spines, and other forms of protective armor and some into teeth and claws. Skeletons provided body armor for prey and ways of breaking, cracking, or drilling through body armor for predators. A skeletal framework for the body also provided locations for muscle attachment and thus greater body mobility. With a firm body structure, muscles could pull against something solid rather than pulling against each other. Heads, gills, and digestive tracts could be stabilized within a fixed framework as fins and legs moved the body forward. Not all Cambrian animals developed skeletons (in fact, only about a third did, as best we know), but the ones that did are the ones that left the best fossils. Untold numbers of species and phyla that roamed the Cambrian oceans left no fossil record at all, and we know nothing of them.

Mineral skeletons are an important development for another reason. They are an organic and integral part of the animals that have evolved them, yet they are not themselves alive. They do not eat, breathe, or reproduce; yet the organism is what it is because of them. It is easy for us to see this; I mention it only because it is not so easy for us to see and understand non-living elements of our own lives. Perhaps because they are so close to us, we see houses, cars, tools, pipes and wires as non-living and external to ourselves, even though we are what we are because of them. They are not alive and do not eat or breathe, but like the carbonate skeleton of the Cambrian trilobite, they are integral and organic to the form of life that we have become. Interestingly, many of the recently invented non-living components of human civilization are polymers like polyethylene, polystyrene, and polyvinyl chloride that we have learned to synthesize from organic monomers in much the way cells learned to synthesize proteins. Technology has become a sort of exoskeleton that encloses, extends, and mobilizes us in ways that we do not easily see and understand.

Animal phyla are basic body plans that evolved over about a 10 million year period within the Cambrian. (The entire Cambrian period lasted from about 545 to 495 million years ago). About two thirds of the phyla that evolved were extinct by the end of the period. No new phyla have evolved since. About thirty phyla survive today, nine of which are considered major: *porifera* (sponges), *cnidarians* (jellyfish, sea anenomies and hydras), *platyhelminthes* (planaria and other flatworms), *nematodes* (roundworms), *annelids* (earthworms and other segmented worms), *mollusks* (snails, slugs, clams, squid, octopi), *echinoderms* (starfish and sea urchins), *arthropods* (insects, spiders, crabs, lobsters), and *chordates* (fish, amphibians, reptiles, mammals, and birds). All phyla evolved in seawater and four are represented on land: annelids, mollusks, arthropods, and chordates. There are fewer phyla on land than in the sea, but many more *species* on land, as there are far greater ranges of temperature, geology, aridity, and other environmental conditions.

Phyla are divided into classes, classes into orders, and orders into families. The arthropod phylum, for instance, consists of five classes: crustacea, insects, arachnids, centipedes, and millipedes. The chordate phylum (mostly vertebrates) has classes of fish, amphibians, reptiles, mammals, and birds. Orders are groups within classes. Insects, for instance, are divided into orders of beetles, moths and butterflies, ants and bees, flies, dragonflies, etc, while mammals are divided into orders of bats, dolphins and whales, rodents, ungulates, carnivores, and primates. Examples of families within orders are felines (lions, tigers, and house cats), canines (dogs, wolves, foxes), ovines (sheep and goats, etc.), bovines (cows), and equines (horses and zebras). Families are further divided into genera, each of which consists of several species. Tree genera are oaks, willow, elms, etc. Within the oak genus are the species red oak, white oak, live oak, pin oak, blackjack oak, etc. A species is generally defined as a population of individual organisms that interbreed successfully. Species may themselves be divided into varieties (calicoes, Persians, tabbies, Siamese, etc.). New species often arise when varieties evolve in isolation from one another to a point where they can no longer interbreed.

For ten million years, an army of eukaryotic cells marched headlong into the wall of an early Cambrian arms race, bodies piling into tissues of claw, leg, tentacle and digestive tract. Safety was in organization and numbers: some cells sought work in locomotion and defense, others in communication and digestion. Every combination was tried. Each succeeded until it failed. From the turbulence, a higher level of life emerged.

During the evolution of the animal kingdom in the Cambrian period, basic rules were established for all time as to how best to arrange cells for eating, breathing, protecting, and procreating. The force of life tried every idea it could come up with. New ideas on this most fundamental level are no longer possible. New families, genera, and species may evolve, perhaps new orders, but no new phyla. Since the Cambrian Explosion multicellular evolution has been a fine-tuning of living forms within firmly established outlines.

III

Phyla

SPONGES

With a single whip-like flagellum attached to its outer membrane, a cell can make a pretty good living without the help and company of other cells. Free-floating or attached to the substrate, it gathers enough bacteria, plant debris, and other microscopic nutrients with this single appendage to meet its needs without coordinating efforts with neighbors. There are many species of protista (single-celled eukaryotic organisms) known as *flagellates* that survive in this manner as rugged individuals.

But it is a difficult and lonely life, suitable only for the hardy few. Hours are spent every day, waving the long flagellum back and forth, back and forth, without finding a single morsel. The volume of seawater that must be filtered for a few morsels can be overwhelming for so tiny an organism. The flagellum captures food directly, but it also keeps water moving past the cell's gullet to bring food within capturing range. It keeps waving between captures to keep the micro-current in motion. When the current stops, the food supply stops. The current also provides fresh oxygen and carries away carbon dioxide and other wastes. Other flagellate cells in the area can be a problem, as they, too, are busy depleting surrounding waters of nutrients and contaminating them with wastes. Thousands of independent flagellates in the same neighborhood, each creating its own micro crosscurrent, can become a terribly inefficient proposition. If, as an alternative, there were a way to get all the flagella of all the cells to work together to create a single macro-current flowing in a single direction, all cells would have access to a continuous supply of fresh seawater without each working so hard: Hence, the sponge, and incidentally, the animal kingdom.

Sponges belong to *Porifera*, the first and most primitive animal phylum. The name refers to their porous outer skin. To coordinate efforts, flagellates line up and attach themselves to one another. To facilitate water flow, they form a porous, bowl-shaped structure where combined flagellations create many inward water currents through small spaces between the cells. A single outward current flows through the opening of the bowl. The opening is known as an *osculum*. The bowl-shaped enclosure performs for individual sponge cells what we higher animals have come to know as digestion and circulation, and is therefore called a *gastrovascular* cavity. As before, each flagellated cell gathers food for itself. But now, with all cells beating together and more water flowing past, there is considerably more available for each, with less effort. Each cell concentrates more on gathering food and less on keeping the water moving. This happens with no systematic coordination of flagellate activity; what keeps the water flowing in one direction is the alignment of cells in the body wall of the cavity. The system works so well that there is an abundance of food that can be shared with other types of cells in the body of the sponge – cells that do things for the community besides gathering food. There is, in other words, enough food to provide a division of labor. Variations on this central concept have built the cell communities we call animals.

A community of cells needs structure and protection. To provide these, flagellated sponge cells feed and support other cells without flagella. Some, the *Pinacocytes*, line and protect the outside of the bowl. These hexagonally shaped cells form a thin but semi-hard skin-like covering that gives support to the flagellates on the inside. Pinacocytes also have the ability to contract and reduce the size of the bowl when conditions become unfavorable for normal functioning. Those located at the bottom of the bowl attach themselves to rocks, coral, or other substrata, and become the base that stabilizes the community in a single location. Other non-flagellates became tubular macaroni-shaped *Porocytes* at openings in the pinacocyte layer. Seawater passes through porocytes to the interior of the bowl.

They contract and restrict the current when necessary, or cut it off altogether. Other specialized cells, *Myocytes,* develop at the osculum, or bowl opening. These "muscle" cells earn their living by regulating the size of the opening, controlling the jet of water leaving the bowl. This is of great importance, as water laden with wastes and depleted of food and oxygen must be expelled at some distance to keep it from cycling back through the pores. At times of decreased flagellate activity and decreased outflow, myocytes contract, constricting the size of the osculum, creating more pressure, and expelling the outflow to a greater distance. This simple stratagem is so effective in keeping inflow water fresh that myocytes receive a regular supply of food from thankful flagellates.

Flagellates themselves become specialized in sponges. Relieved of the rigors of individual survival, they extract more food from the water than they need for themselves. Lining the inside of the gastrovascular cavity, they feature ciliated "collars" around their gullets through which the flagella directs food particles. Commonly known as "collar cells," they are referred to in the scientific community as *choanocytes,* from the Greek word *choane,* or "funnel." In exchange for the benefits of communal living, choanocytes pass on much of the food they gather to the other cells. They partially digest some of this food before passing it on.

Another type of cell, the free-moving *archeocyte,* accepts food from the collar cells and distributes it to the rest of the community. Archeocytes (alternatively known as mesenchyme cells or amebocytes) move about like amoeba between the inner and outer layers of the bowl. They do more than distribute food. They can divide rapidly and differentiate into any other type of sponge cell: pinacocyte, porocyte, myocyte, or choanocyte, as needed. This feature distinguishes the sponge from higher animals. Sponges do not have true "tissues" in that their cells are not always born and raised in a particular profession. Cells can change jobs in mid-career. Sponge cells of a particular type can divide and replace themselves with new cells of the same type, but archeocytes often step in and provide new growth in a variety of cell types as the community expands. They are the "stem cells" that seem to know what type of new cell is needed where and when.

They are also instrumental in reproduction of the sponge community as a whole. In the sexual reproduction of sponges, it is the archeocytes and choanocytes that divide (by meiosis) into either sperm or egg cells. Some sponge communities are male or female, while others are hermaphrodites (providing both male and female sex cells.) In hermaphroditic sponges, sperm and egg cells are usually released at different times to prevent self-fertilization. Sperm are released into the cavity and sent out through the osculum, often in mass, and find their way through the pores of other sponge bodies, where fertilization takes place. Zygotes (fertilized egg cells genetically distinct from either parent organism) are incubated in the body cavity and released (usually through the osculum) as flagellated larvae. Larvae swim or crawl a short distance from the parent, become attached to the substrate, and grow into a new sponge.

Archeocyctes are also involved in asexual reproduction. Under conditions where sponges are unlikely to survive, such as drought or winter weather, they often form *gemmules* before they die. These are small bundles of archeocyctes surrounded with food and a protective outer layer that can withstand desiccation or freezing. Gemmules remain intact after sponge bodies die and disintegrate. When conditions improve, the gemmule opens and archeocyctes emerge through a pore. They collect in a formless mass on the substrate, then divide and differentiate, shaping themselves into a new organism. The new sponge body is genetically the same as the parent. Archeocytes are involved with other types of cells in two other forms of asexual reproduction. In *budding*, a small gastrovascular cavity begins to develop and branches from a mature cavity. It can remain attached to the parent organism and set up shop as a separate gastrovascular operation, or it may break free. If it breaks off, it will drift through the current and settle on the bottom, becoming a separate organism. In *regeneration*, almost any piece of tissue torn from a sponge body will, under favorable conditions, grow into a complete new organism.

Sponge cells are differentiated into specialized functions, but each performs all life-functions, and can survive to some extent on its own. There is exchange of partially digested food between cells, but there is no extracellular enzyme secretion. All digestion is intracellular: The sponge as a whole can eat nothing that cannot be eaten by a single cell. There are no specialized sensory cells. There are no nerve cells connecting one part of the sponge body to another, and no neurologically coordinated movements. Coordinated movements, such as they are, seem to be initiated by simple "word of mouth" messages between adjoining cells. A disturbance may lead to a contraction of pinacocytes in the immediate area, but no general contraction though the rest of the bowl. There is no unified response of the organism as a whole. Contractions may spread gradually across the body from one cell to another, but there is no message sent to other parts of the body, and no way for one part of the body to "know" what is happening elsewhere. Coordinated motion of flagella seems to be built into the structure of sponge itself, rather than in response to stimuli (flagella always conduct water in the same direction), and contraction of myocyctes at the osculum is probably a response to a locally detected drop in water current, rather than to any communicated drop in flagellate activity. The signal to constrict the osculum is sent by the choanocytes to the myocytes through changes in water currents, by design or by accident, and not through any stimulus within the sponge body itself.

Sponges are the only animals with no mouth, taking in food through many pores and excreting through the bowl opening. Only the most primitive sponges remain at the level of a single, unfolded gastrovascular cavity. Higher species increase their overall size by branching and folding the cavity into many connected chambers. Folding of the cavity maintains maximum surface exposure for choanocytes. To give shape and support to the soft tissue, sponges secrete hard calcareous or siliceous *spicules* throughout the walls of the chamber. It is the many-chambered sponge skeleton, washed of its living components, that is the natural bath sponge available to consumers in specialty markets.

The bowl-shaped gastro-vascular cavity is the flagellate's answer to the problem of coordinated activity, and the osculum is its way of avoiding the double work of re-processing water it has already seen. But in making life easier for itself, the single cell has lost the freedom it once had. This raises the question of *individuality* in the sponge community. Where is the *being* of the sponge? Is it in each cell or in the community as a whole? Is the sponge a cell colony, or is it an organism? There can be little doubt that sponges have evolved from single-celled flagellates, but at what point do individual cells lose their separate identities and become an organism? Is there any sort of consciousness associated with the organism as a whole? There is a wholeness arising from cell communities that supersedes the individual cells of which it is composed – a distinct *being* of multicellular organisms? Where does it begin and end?

A well-known experiment in sponge regeneration illustrates the problem of individuality. A full-grown living sponge is passed through a fine cloth filter and pulverized to individual cells or small groups of cells. These are then placed in seawater. Almost any higher organism would, of course, be killed in this process, cells and all. But sponge cells not only survive, they move about on their own until they find other cells, and then re-aggregate into a mass. Choanocytes migrate toward the middle of the mass, pinacocytes toward the outside, and archeocycts toward the space in between. Soon the mass becomes attached to the substrate, hollows, begins filtering water, and forms a gastrovascular chamber and an osculum. Is this a new organism, or the same one? Did the original sponge "die" and a new sponge "come to life?" Can two distinct organisms consist of the same cells? A variation of this experiment has been performed in which sponges of two separate species of distinct pigmentation were disassociated through a cloth filter and placed together in seawater. The cells moved about, bumping off each other as before, but aggregated only with other cells of their own species.[5] There seems to be some species-specific means by which sponge cells recall and reestablish their former community.

The question of individuality is further complicated by the tendency of advanced sponges to form more than one gastrovascular cavity per osculum, and to cover more than one adjoining osculum with a layer of pinacocytes. A number of cavities often use a single opening, and a continuous outer layer of cells often covers several openings. Most scientists are willing to give sponges the status of organism (rather than cell colony), but are then stuck with the question as to what constitutes the organism: A gas-trovascular cavity? A group of adjoining cavities with the same osculum? A group of cavities and oscula covered by a continuous layer of pinacocytes? Some give the individ-uality prize to the gastrovascular cavity, as it is here that cells are truly working together and truly interdependent; others give individually to a group of connected cavities, for the same reason, in that the workings of the osculum bind them to a common dependency. Some scientists claim that groups of adjoining cavities are merely colonies of in-dividual sponges and that cells in separate cavities have no functional relation to one another. Others point out that ar-cheocytes, even those involved in reproduction, move freely from one chamber to another in large complex sponge colo-nies with multiple chambers and oscula. A sponge growth covered by a continuous layer of pinacocyte cells consti-tutes, for them, the individual. They claim that in some sponge species, *Verongia gigantean* and *Tethya crypta* in par-ticular, there are distinct rhythms of osular expansion and contraction in response to purely endogenous (internally induced) stimuli, which is a clear attribute of individuality.[6]

Questions of individuality are more about us than about Porifera. Sponges don't give much thought to the problem. We consider each of ourselves a distinct consciousness, and want to know where a being begins and ends, and with what, exactly, we may identify as a fellow sentient. We want a clear unit of selfhood to which we may in some way re-late. For the sponge, it is clear that there is no such distinct unit of consciousness. But neither can there be for any animal, in any absolute sense, as we are each a composite of individual cells – a multi-cellular colony – however highly integrated. Separate individual consciousness within an organism may be something we impose from the outside.

Hydras

Hydras belong to the next major animal phylum, the *Cnidaria*, which includes jellyfish, sea fans, sea anemones, and coral. Cnidarians bear some similarities to sponges, but their cell structure is more highly differentiated and specialized, and their body plan begins with an entirely new principle. There are no pores through which seawater enters the main body cavity. Instead, water and food enter and exit through a single large opening at one end, the *mouth*. Sponges do not have mouths. Cnidarians and all higher animals have the same fundamental body plan: a digestive cavity connected to the outside world through this opening. (Proboscis worms and round worms improve on this basic body plan with a second opening, or *anus*, and one-way traffic through an *intestine* connecting the two openings.) Because of this fundamental difference, most zoologists believe that cnidarians and all higher animal phyla evolved from a separate line of flagellate protozoa and not from Porifera.

The hydra is a small aquatic animal, about a half-inch long, consisting of a digestive cavity with stinger-equipped tentacles at the mouth end. Like the sponge, it has amazing powers of regeneration – broken pieces often grow into complete new organisms – and it is from this ability that the hydra gets its name. *Hydra* was a nine-headed monster of Greek mythology that grew back two new heads for every one that was cut off.

Where the sponge is restricted to feeding on microscopic particles that can be ingested by a single cell, hydras, because they do not ingest food through pores, are capable of feeding on much larger prey. Attached to the substrate, they do not chase or hunt prey, but wait for passing worms, crustaceans or other small animals, their long tentacles drifting about in the current. Special stinging cells inject poisons once prey is encountered, killing it or numbing it for long enough to bring it in through the mouth to the gastrovascular cavity. Unlike sponges, hydras have a *nerve net* linking cells throughout the body, so that when one tentacle senses prey, others are alerted and slowly close in on it. Tentacles work together to entangle and disable the prey as the mouth opens to ingest it. Hydras have special *gland cells* that secrete digestive enzymes once the prey is engulfed. This extracellular digestion breaks food down into pieces small enough to be assimilated by individual cells. When digestion is complete, the mouth reopens and waste products are expelled.

Cnidarians are characterized by two distinct layers of cell tissue. The outer layer (ectoderm) consists of protective cells with muscle fibers running lengthwise. When these cells contract in unison, the body of the animal shortens. When cells on one side contract, the body (or tentacle) bends to that side. Cells of the inner layer (endoderm) lining the gastrovascular cavity are specialized primarily for digestion, but they, too, have muscle fibers. Their muscle fibers run in a circular pattern around the cavity, at right angles to those of the outer lining. When they contract, the cavity tightens and lengthens. With these two parameters of cellular coordination, the hydra is capable of a wide range of body motions. Two cell layers are present in the sponge, but in the hydra the inner and outer layers are more specialized, include more different kinds of cells, and are capable of highly coordinated activity, thanks to the *nerve net*.

A tightly coordinated group of cells performing a specific function is known as a *tissue*. Sponges display a degree of tissue development, but it is only with the nerve net of the hydra that cells can truly work together. Cnidarians are, in fact, the first animals to show the major tissue types of all animals: *epithelial* (the two layers that in higher animals become skin, intestine lining, blood vessels, and glands), *muscular*, *connective* (that become bones, blood, and mesentery tissue), *reproductive*, and *nervous*. Higher animals have many more types of cells than the hydra, and many more types of tissues that themselves evolve into *organs* and *organ systems* that the hydra does not have, but all of these are modifications of fundamental tissue cells that the hydra does have.

Sponge cells communicate through adjacent physical contact, while hydra cells keep in touch through the nerve net. Special sensory cells imbedded in both the inner and outer layers detect chemical changes and contact with external objects. The nerve net acts as an intercellular medium to keep other cells informed, in some encoded way, of what the sensory cells are experiencing. The nerve net also coordinates activity: Cells are stimulated simultaneously and know when to contract or secrete in unison. But there are no ganglia and no central control systems in the hydra; it has nothing approaching a brain. There is a slight concentration of nerve cells at the mouth and base, but no central location where nerve impulses are differentiated and evaluated. There are *synapses,* or spaces between nerve endings over which signals must pass, as in higher animals, but in the hydra signals pass *both ways* across a synapse.[7] This means that, unlike in a brain, there can be no definite pathway for a signal to follow, and therefore no particular pattern of nervous impulses. Only very basic signals pass through the net, and they seem to go everywhere at once. A nerve impulse seems to be interpreted according to the type of cell receiving it: if you are a cell with muscle fiber, "contract now!" – if you are a cell with enzymes or poisons, "secrete now!" This one-dimensional response pattern greatly limits behavioral possibilities. A strong stimulus anywhere will simply cause the entire body to respond. If you touch the side of the hydra's body, the entire animal will contract into a tight ball.

The hydra is, however, capable of limited locomotion. For the most part it remains attached at its base to the substrate with tentacles drifting overhead. But if it has not found food for some time it may move in a "somersaulting" motion. The body bends to one side until tentacles at the top end touch the substrate. When enough tentacles have contacted to temporarily support the rest of the body, it lets loose at the base and passes upside-down over the tentacles. When the base re-contacts the substrate, the hydra body re-rights itself and straightens to a right-side-up position. Tentacles regain their overhead position and resume the search for food.

The problem of individuality is not so severe with the hydra. Like the sponge, the hydra is capable of regeneration and budding, which muddies the clear-cut organic unity we are used to in the case of higher animals, but the individual cells of the hydra are so much more integrated into the wholeness of the organism that we are not tempted to call them a colony. The nerve net, coordinated muscular movement, and the ingestion of food much larger than can be assimilated by an individual cell make the wholeness of the hydra more than the sum of its cells.

PLANARIA

The darkly colored *planaria*, a member of the "flatworm" phylum *Platyhelminthes*, is a half-inch to an inch long and inhabits inland streams and ponds. It is characterized anatomically by a third layer of cells (mesoderm) between the outer and inner layers (ectoderm and endoderm) that we have already seen in the cnidarians. It is from this middle layer that the muscles, bones, and other advanced organs of higher animals develop. (In cnidarians there is no distinct muscle cell tissue; muscles fibers are present only within cells of the inner and outer body layers.)

The planaria has distinct head and tail (anterior and posterior) ends and also front and back (ventral and dorsal) surfaces. The body plan has a distinct directional orientation and *bi-lateral symmetry*. There is only one way to cut it and have two equal halves. All higher animals including humans have adopted the bilaterally symmetrical body plan: To cut a man into equal halves, for instance, you cannot cut his top from his bottom or his front from his back – you have to cut his left from his right. You must slice him right between the eyes to be sure and get one ear, one arm, one leg, etc in each half. This symmetrical arrangement begins with the flatworm. The hydra, on the other hand, has *radial* symmetry. It is rounded, like a cylinder, with tentacles going out in all directions and no distinguishing features in any. You can cut him anywhere along his central axis and have two equal halves.

Like other bilaterally symmetrical animals, the planaria is built for forward motion. Its relatively long and thin shape allows for easy passage through water, and sensory organs at its anterior end give it some idea of what lies on the road ahead. Also at the head end are two tiny eyespots and a concentration of nerve ganglia that some zoologists are brave enough to call a *brain*. It does not swim freely through the water, but slithers along the bottom, usually wagging its head side to side in search of food, or in search of a chemical gradient that may lead to food. But the mouth, interestingly, is not at the head. It is instead located about halfway down the body on the under surface. When the worm encounters a bit of animal matter or a small live animal it slides its flat body over the prey, trapping it against the bottom. A long, flexible *pharynx* extends from the mouth and injects enzymes into the prey. Pieces are broken off and swallowed into the gastrovascular cavity, and assimilated there by the cell lining. From the cells of the cavity lining food particles (and oxygen) are distributed to all parts of the body. There is no separate circulatory system.

The lack of a circulatory system creates a plumbing problem for flatworms. Hydras and sponges have only two cells layers so that each cell has direct contact with water from either the outside or the gastrovascular cavity. Each cell is constantly bathed in a fresh supply of dissolved oxygen, and can get rid of carbon dioxide and nitrogenous wastes on its own. But the middle layer of cells in the flatworm are cut off from direct contact with water, and require a systematic means of disposing of wastes. Apparently, the biggest problem faced by the planaria is regulating the water and salt content of its muscle tissues between the inner and outer layers. A network of fine tubules running the length of the endoderm connects these cells to the outside. Ciliated *flame cells* distributed throughout the middle body layer are connected to the tubules and create micro currents that gather wastewater into them. The wastes are then conducted through the tubules and out of the body through special excretory pores. The flatworm body is small and thin enough for most of the gaseous metabolic wastes (carbon dioxide and nitrogen) from its middle cells to be diffused through the other cell layers without excretory tubules. Plumbing systems similar to that which the planaria has developed for water regulation become a necessity in larger bodied animals for all types of waste removal.

The planaria, like the hydra, has both longitudinal and circular muscles and a nerve net to coordinate them. But it also has a *central nervous system*, with a *cerebral ganglion* (if not a full-fledged brain), two parallel nerve cords running the length of the body, and numerous sensory cells and sensory organs. The ganglion and most of the sensory organs are concentrated in the head. Unlike in the nerve net, nervous impulses travel in one direction across synapses. This makes distinct pathways possible and the potential for differentiated behavior. The ganglion receives impulses from sense organs, evaluates them, and returns a selected response to body muscles. We will not call this thinking, but it is without doubt what thinking starts out to be.

It is interesting that while a central nervous system is necessary for selected responses, it does not appear to be necessary for muscle coordination. If you cut the ganglion out of a planaria, its muscle cells, with the help of the nerve net, will continue to move in unison. Without its brain it remains able to do things, but not anything in particular.

The "retinas" of planarian eyespots consist of a handful of cells in a pigment cup that lets light in from only one direction. It is doubtful that the planaria can resolve any sort of image with its eyes, as they have no cornea or lens. As planaria avoid light, and regularly change their direction away from it, the eyes probably do little more than inform the animal of its orientation to a light source. More important than the eyes are sensory lobes on either side of the head that inform the cerebral ganglion of chemicals in the water. A chemical gradient indicating the direction of an object of prey (or of a predator) can be determined by comparing stimuli on one side of the head to that on the other. Planaria augment this gradient sensing ability by moving their heads from side to side, or by moving in one direction for a distance and then in another, testing the waters as they go. This is how they find what they eat and avoid what eats them. Creating a sense of gradient by comparing chemical perceptions and selecting appropriate muscular responses is probably most of what the ganglion does.

The eyespots are not the windows to the soul that they are in higher animals. The sensory lobes on either side of the head, however, look like ears and along with the eyespots give this particular flatworm something that looks like a face. You could almost say hello. Here, at last, we have come to a fully recognizable multicellular "unit of consciousness:" the first sentient wholeness more than the sum of its cellular parts. If so, there remains a problem, and an eerie one at that: Like sponges and hydra, planaria are capable of regeneration. Cut one in two and each part will grow back into a complete, separate organism. Worse, if instead of cutting a piece off, you slice up the middle of the head right between the eyes just a little way, each half will grow back to a complete head, "brain" and all. One organism, two faces.

Relativity and the Cuttlefish

The cuttlefish is not a fish, but a relative of the octopus and member of the phylum *Mollusca*,[8] and class cephalopoda. About a foot long, it has no legs, no backbone, and no skeleton. What it does have is the most advanced sensory organs outside of the Chordate (vertebrate) phylum. It has large, vertebrate-class eyes that are true windows to the soul. If you look at him, he will look back.

Most interestingly, the cuttlefish is physiologically capable of understanding Einstein's Theory of General Relativity. This is due to his possession of a special sensory organ called a *statocyst* by which he is aware at all times of the direction of gravity. The statocyst is a small fluid-filled organ that helps the cuttlefish maintain body stability at night or in turbulent ocean waters deep below the surface where there is little light. It helps him know which way is up at times when there may be no other way of knowing. But it does more than this. The cuttlefish, both predator and prey, must be ready to move rapidly in any direction. It does this by ejecting a jet of water from a siphon that emerges from the main body muscle (the *mantle*). In moments of fight or flight the cuttlefish squirts himself up or down, right or left, as the occasion requires. The statocyst, because it is sensitive to accelerations as well as to gravity, helps him keep track of where he is going. But acceleration is not the same thing as velocity: The cuttlefish may be swimming along at any velocity, even an extremely high velocity, and the statocyst will not know it. Acceleration – what the statocyst does know – is a *change* in velocity: an abrupt switch from zero to five meters per second or from five to ten. Acceleration can also be from right to left or from up to down. (Velocity has two components: speed and direction; a change in either is a change in velocity. Acceleration, therefore, may be a shift in direction when speed remains constant. Slowing or coming to a stop – a *de*celeration – is also a change in velocity, and a form of acceleration.) Any sudden shift in motion will cause fluid in the statocyst to slosh back or forth, right or left, and notify the cuttlefish that he is accelerating. This gives him a first-hand experience of Einstein's *Principle of Equivalence* between gravity and acceleration, the cornerstone of his Theory of General Relativity.

Einstein showed that gravity and acceleration are one and the same. In a space ship with curtains drawn over the windows you would not be able to tell whether the ship is standing still in a gravitational field or accelerating through interstellar space. If it were standing on the surface of a planet you would be pulled to the floor of the ship by gravity; if in space, you would be pulled to the floor away from the direction of the ship's acceleration, and it would feel the same as gravity. You would feel a "downward" force either way, and watch all unsupported objects in the ship "fall" to the floor. As long as you do not look out the window and see whether or not things are flying past, there is no way to know the difference. (The velocity of the ship makes no difference, only the acceleration. You cannot feel a million or a billion meters per second of velocity, but you *can* feel as little as one meter per second *per second* of acceleration.) There are no physical experiments that you can perform to distinguish between acceleration and gravity because they are not physically different. They are the same thing. This was an enormous breakthrough in human understanding of the physical universe. The cuttlefish already knew it.

Cuttlefish are not the only ones. Many quite primitive animals have statocysts. Clams have them. Animals as "low" as the Cnidarian jellyfish have simple statocysts that consist of no more than a cup of cells with a grain of sand bouncing up or down or back or forth as the animal moves about. Just about any animal capable of controlled movement has a statocyst of some kind and therefore should have an intuitive understanding of the Principle of Equivalence. We humans have statocysts in the form of semi-circular canals in the middle ear.

So what's special about the cuttlefish and what's the big deal with Einstein? The big deal is in reconciling what we *feel* with what we *see*. Statocysts constantly remind us that acceleration and gravity are the same, but that's not what it looks like. Acceleration and gravity might feel the same, but as long as I am not bottled up in a spaceship with the curtains drawn, acceleration looks like rushing through space at greater and greater speeds, and gravity looks like standing still. I can tell the difference by what I *see* when I push the spaceship curtains back and look out the window.

So can the cuttlefish. He is the not the first to have eyes, but he is the first of the invertebrates to have eyes pretty much like ours. The class of mollusks to which he belongs, the *cephalopods* (squids, nautiluses, cuttlefish and octopuses) is the first to have uncompounded eyes with not only a cornea and lens, but lots of retinal cells for a high degree of visual acuity. If not the first to look out the window of the rocket ship, he is the first to know what to make of what he sees. The structure of his eyes indicates that he should resolve clear detailed images of objects around him and determine relations between objects *in space*. He is probably the first invertebrate to have an understanding of three-dimensional space similar to our own, the first to grapple with the difference between what he sees and what he feels and, therefore, the first to be *surprised* by the Principle of Equivalence and its implications in the Theory of General Relativity.

Now let me run through that again. The statocyst is an organ through which many animals gauge their orientation to gravity and acceleration. When an animal is loafing around between meals, the gauge remains stable in a familiar direction and intensity. When he starts to move, it changes direction and intensity. But it does not tell him anything about space. It cannot even tell him whether or not he is moving – only if he is *changing velocity* (speed or direction). With the statocyst *alone* he might define directions "up" or "down" or "right" or "left," and identify "down" with that loafing around feeling, but he can have no detailed sense of spatial relations among objects in his environment. There is no window in his space ship, so the principle of equivalence is intuitively obvious. It makes sense because, with the statocyst alone, he has *no concept of space.*

When he develops a good set of image-resolving eyes, things get complicated. He develops a sense of space, but it does not jive immediately with that gut feeling in the statocyst. Simple, unaccelerated motion in space does not activate the statocyst. But he must reconcile what he sees with what he feels (and touches, hears, and tastes) in order to develop a concise and unified picture of physical reality. All of his perceptual realms of consciousness have to make sense to each other, and he has to come up with a way to coordinate them in space and time. He coordinates *seeing* with *feeling* by equating his visual experience of *changing* velocity – not velocity itself – with his sensation in the statocyst. When he sees objects flying by at greater and greater velocities he feels fluid shifting back in a canal. The direction of acceleration is coordinated with the location of the sensation in the statocyst, and the rate of acceleration is proportional to its intensity. That is how seeing coordinates with feeling, and where the cuttlefish gets a sense of "body in space." But it is not really the body in space. It is the body in *space-time*. What he feels in the statocyst is his body's relation to the whole of space-time. When he moves in relation to it by accelerating, or by lying stationary in the Earth's gravity, the statocyst sounds off. The coordination between seeing and feeling works well, but somewhere over the millenia the cuttlefish loses the ability to distinguish between space and space-time, and forgets that dimensions are there to do the coordinating. He comes to suppose instead, in a more commonsense way, that only space really exists and there is an inexplicable property of space that causes his body to "drag" in the second time dimension (the per second – *per second* of acceleration). Later phyla have come to call this phenomenon "inertia."

This is a more simply understood way to coordinate seeing and feeling for everyday siphoning about the ocean depths. It is more relaxing to think of the universe in terms of up-down, right-left, and back-forth, with a disassociated past-present. The business of a second time dimension associated with gravity, inertia, acceleration, and sensation in the statocyst was shoved somewhere into the first time dimension. The only drawback to the simplified view is that, in forgetting the difference between space and space-time (and space-time-time), the cuttlefish no longer intuitively visualizes the principle of equivalence. Gravity no longer looks like acceleration. A direct, intuitive understanding of general relativity was not, apparently, as important to him as a simplified notion of body in space, and he decided to go with simplification. The Ordovician job market being what it was, he became a hunter. He (and the rest of the animal kingdom) forgot about the principle of equivalence. Einstein brought it back several hundred million years and a phylum or two later.

With greater access to theories of electromagnetic physics, and additional well-placed ganglia, Einstein restored the Principle of Equivalence that the cuttlefish left behind. He did this by rediscovering our location in space-time, reminding us that there is another dimension that we have to throw into the mix. Not only that, he showed that space-time can be *curved*. Four-dimensional space-time is curved into yet another dimension. (Einstein didn't say this part – I did.) *What we feel in the statocyst is the curvature of space-time.* It kicks in when we curve "flat" space-time by accelerating, or get curved in "curved" space-time by remaining stationary in gravity. The value of the "per second-per second" dimension is proportional to the intensity of what we feel. Acceleration is "our own" curvature of space-time that makes objects move by us faster and faster, and gravity is a curvature of space-time itself near massive objects that makes objects move toward them faster and faster. When you look out the window of the space ship and see that it is not accelerating through interstellar space, but just sitting on the ground, all the houses and trees and baseballs *are*, in fact, accelerating past you at exactly the rate you feel in your statocyst. They don't move down because the ground supports them, the same way you and the space ship are supported. In a gravitational field without ground to hold you up, you and the spaceship and the houses and trees would all "fall" at the same accelerated rate. Until you hit the ground, you would be stationary relative to them, there would be no curvature, and your statocyst would measure zero.

If we could get away with saying that the Theory of General Relativity is just an elaboration of the Principle of Equivalence, we could also get away with saying the cuttlefish has the hardware, if not the software, to comprehend it. We might even get away with saying that what Einstein accomplished was to show us the dimensional relation between the tactile and visual realms of consciousness.

Ants

Ants are an insect family of the phylum[9] *Arthropoda*, whose members are distinguished by a hard exterior body covering, or *exoskeleton,* and jointed legs. Other arthropods include spiders, centipedes, millipedes, crabs, shrimp, and lobsters. If biological success is measured by number of living species, arthropods win the prize hands down, with over a million species known to science and an estimated 3 to 100 million species as yet undescribed.[10] All other animal phyla combined amount to around 250,000 known species. Among classes within the arthropod phylum, *Insecta* wins the diversity prize with about 90% of the species, and 7000 new species described every year. Insects win in terms of shear numbers, too, with an estimated billion billion (1,000,000,000,000,000,000) individuals on Earth. That's a lot of bugs: two hundred million ants, bees, butterflies, mosquitoes and cockroaches for every human. (We may console ourselves with prizes in other categories, even if we have to make them up.)

The arthropod body plan is an enormous departure from that of earlier animals, and accounts for much of the phylum's success. The exoskeleton provides much improved protection from predators and, more importantly, a body frame for muscle attachment. Muscle fiber in earlier animals serves mostly to move tissue *within* the organism. Locomotion as such requires contortions of the body as a whole, digestive tract and all. A rigid arthropod skeleton, on the other hand, provides stability for body organs not involved in locomotion. Muscle fibers move jointed appendages without contorting the gastrointestinal tract, and much more rapid and purposeful locomotion becomes possible. Arthropods also secrete a waxy layer on the exoskeleton that prevents water loss, thereby preparing the arthropod body for emergence from the sea. The exoskeleton body plan, originating in seawater with crustaceans (lobster, crabs, shrimp, etc) becomes so well adapted to life on land that it leads to a veritable explosion of terrestrial forms. Arthropods are the first animals to invade the dry land, arriving some 400 million years ago (after an earlier assault by members of the plant kingdom). On land and in the air above land, wave upon wave of insects has evolved and taken the Earth.

With the exception of the exoskeleton, the insect body plan is similar to that of segmented worms (phylum *Annelida*). Larval development begins with nineteen body segments that become three basic body parts in the adult. The first five segments become the *head*, the middle three the *thorax,* and the posterior eleven the *abdomen.* The mouth, feeding appendages, sensory organs including simple and compound eyes, and brain are concentrated in the head. (In addition to the head ganglia that compose the brain, each segment maintains its own nerve ganglion for internal control.) The walking appendages (always six) and wings (if present) are attached to the thorax: a pair of legs for each of the three segments and a pair of wings for the second and third. Digestion, respiration, excretion, and reproduction occur mainly in the abdomen. Circulation is quite different from segmented worms because of the exoskeleton. Where annelids developed a *closed* circulatory system, (similar to vertebrates) with blood flowing out of the heart through arteries and back through veins, insects and other arthropods have an *open* system, with blood flowing out of the heart, through the aorta, and into the open body cavity. There are no closed arteries beyond the aorta. Because the exoskeleton does not flex, blood pressure developed by the heart (and by general body movement) is not absorbed by the body wall, as it would be in soft-bodied animals. Blood moves over and through body organs from the head back, carrying nutrients, hormones, and waste products, and re-enters the heart through openings toward the rear of the abdomen. Long, hollow filaments streaming through the body cavity known as *Malpighian tubules* remove nitrogenous wastes from the blood while leaving as much water as possible in the body. Rather than ducting directly to the outside, which would be impractical through the exoskeleton, tubules are attached at one end toward the posterior of the digestive tract, so that nitrogenous wastes are disposed in solid form through the anus, in combination with digestive wastes. Blood does not deliver oxygen or remove carbon dioxide. Insects have no lungs and internal tissues must exchange gases with the air directly. Air is brought in and out of the body through pores (spiracles) along the sides of the abdomen and delivered to internal organs through a system of branched air passages (tracheae).

The glaring drawback to the exoskeleton body plan is size limitation. Arthropods cannot grow by simply getting bigger – their hard outer surface is hard from the inside, too, and there is no room to put on new growth. The best answer to this problem is *molting*, or periodically shedding the exoskeleton entirely, and quickly secreting a new one. But there are serious limitations to this stratagem. It is wasteful of body material, the organism is left unprotected as molting proceeds, and very little increase in body volume can be accomplished in any case. So arthropods, insects in particular, remain small compared to vertebrates. Another factor limiting the size of insects is their failure to use their blood for gas exchange. The stratagem of bringing air directly to body tissue through tracheae is suitable only for small body size and would become unworkable for an animal the size of a dog or a horse.

Limitations in the arthropod body plan have diverted evolutionary pressures in an interesting new direction. Rather than making bigger and better insects, evolution has, in the last 100 million years, developed bigger and better organizations among insects. Within the insect class there are more than a dozen orders, including *diptera* (flies, dragonflies, termites, beetles), *lepidoptera* (moths and butterflies), and *hymenoptera* (wasps, ants, and bees.) Of these, only termites and some hymenoptera have evolved in the direction of social organization, that is, of communal living patterns in which the organizational whole becomes of greater survival importance than the individual organisms of which it is composed. Termites and the social hymenoptera exist only as parts of a larger whole; they do not, and cannot, exist as individuals. The colonies within which they live are of so high a degree of physiological and reproductive interdependence that they are often referred to as *superorganisms*. It is the colony as a whole that lives, breathes, moves, perceives, eats, digests, eliminates, defends, and reproduces.

Among the order *Hymenoptera* are the wasp, bee, and ant *families*. Only certain wasps and bees are social, while all ants are social. Within the ant family are seven subfamilies, including *Dorylinae*, the army ants. There are five principle genera of army ants, three in the Americas, one extending into the southern portions of the United States, and two in Africa and Asia. It is the new world genus *Eciton* that interests us most, and within it the species *Burchelli*, which inhabits Central America from Panama to southern Mexico and portions of the Amazon basin. It is most interesting because not only is it social, it is mobile. Most insect colonies are stationary, rooted like plants to nests in trees, rotting logs, or underground, and move only if forced by emergency. Army ants move as a way of life. They hunt in collective swarms of 100,000 or more, and nest in temporary *bivouacs* made of their own bodies. Colonies of *Eciton burchelli* consist of up to one million ants: One queen, two or three thousand males (during occasional sexual broods only), and the rest infertile female workers. Among the workers are four or five *castes* with distinctive body sizes and shapes that perform specific functions: soldiers, raiders, porters, brood tenders, and a special guard unit that remains with the queen, in and out of the nest.

Burchelli ants cycle between a *nomadic* phase in which the colony moves every night for about three weeks, and a *statary* phase, in which the colony remains stationary for about the same or a slightly longer period. During the nomadic phase two or three raiding swarms are sent in separate directions every morning. Each raid begins with a single column emerging from the bivouac that fans out as it nears the hunting ground and forms a swarm. The column may be ten or twenty centimeters wide and over a hundred meters long before it branches into smaller trails leading to various locations in the swarm. The swarm is usually about a meter or two deep and covers a front of from ten to fifteen meters, moving at about thirty centimeters a minute. Nothing is safe in its path. Army ants are exclusively carnivorous. They are not interested in vegetation. Their strategy is to flush out, rustle up, and overwhelm spiders, centipedes, roaches, flies, beetles, small frogs, birds, and snakes, and especially, other kinds of ants. Larger animals can usually move to safety, but any who happen to be immobilized for any reason are in trouble. They do not bother people, at least those who can move faster than thirty centimeters per minute, and they are sometimes welcome in huts and buildings where they do a clean job of exterminating roaches, spiders, and ant nests. It is not wise to resist them. If you hear them coming, you may want to close a few jars and seal away your meat supply. Anywhere near a roving colony of *Eciton burchelli* you will want to be careful about keeping animals in cages.

Individual ants within the swarm move in what appears to be a totally chaotic manner. Each individual moves forward, backward, and side-to-side, picking up and putting down bits of collected booty, bumping against other ants doing more or less the same thing. Yet overall behavior is statistically orderly: The swarm as a whole moves one way or another. All ants leave chemical scents or *pheromones* to mark where they have been, so they know when they have wandered off the trail or away from the swarm.

Returning ants react immediately as they reach the edge of the swarm. Individual ants have to explore new territory, but they always know they are doing so as soon as they no longer sense the pheromone particular to their colony. Wandering beyond the pheromone is dangerous but necessary work – it is beyond the limits of the superorganism and ants that do so proceed in a careful and measured fashion. The pace is slower and antennae are extended to the ground, feeling, tasting and smelling for traces of food or danger. Ants encountering pheromones of other ant colonies or other types of insects report back to the nest. If food is discovered, a piece of it is brought back into the swarm to alert others. Often, smell is enough to bring help. If it finds nothing interesting, an ant will leave a trail of pheromone back into the swarm so that others will be able to precede at least this far without having to stop. If an ant encounters resistance in the form of a larger insect or ant of another colony defending its nest, it releases an alert signal. If it is killed (and it often is), its death will release a signal that will be transferred back along the column to the soldiers.

Soldiers do not engage in foraging raids. The largest in body size of the worker castes, their mandibles (claws) are so specialized for fighting that they cannot even help bring booty back to the nest. They often loaf around the nest or along the trails, usually getting in the way of traffic. But when called to action, they respond quickly and aggressively, without regard to their own safety. They fight to the finish against any challenge and often die in combat, slashing claws, grabbing enemy combatants by the legs and crushing them with huge mandibles, or squirting them with noxious chemicals. In terms of brood care and food allowance, soldiers are the most costly caste for the colony to breed and maintain – and the least productive in terms of direct income – but they pay their way by providing safe passage for their sister workers. They are only about two per-cent of the colony's total population.

The swarm moves only as fast as individual workers are able to explore and chemically treat new territory. Workers within the swarm will not cross the chemical barrier in mass, and thus remain contained in a single teaming horde, crawling, slashing, and stinging its way across the forest floor, over rocks, vines, and logs and into the tree branches overhead. As more ants arrive from the column and through the fan, pressure develops at the base of the swarm, forcing it to surge ahead and outward to left and right flanks. A flattened front develops, several times wider across than it is deep, moving steadily forward in a fairly well defined direction. A cross current of ants marching right and left then develops behind the front line, connecting the flanks and communicating conditions throughout the swarm as it presses forward. Every fifteen minutes or so the entire swarm pivots right or left in a coordinated flanking movement, first charging on one side and then on the other, flushing out prey from under rocks and rotting timbers. As a flank slows, it consolidates its gains, capturing and killing prey, and organizing portaging parties for carrying booty back to the bivouac.

The smell of booty returning to the bivouac and pheromones released by returning workers recruits more workers
toward the location of a successful raid. Ants exploring side
trails sense that something more exciting is going on in that
direction and join the march. A less successful swarm raiding in another direction may abandon its efforts altogether
and begin streaming to where the booty is. Smaller prey
is carried back whole, while larger arthropods and small
vertebrates are torn to pieces. Ants often team up on larger
pieces, carrying them back in a concerted lockstep fashion
that moves more total material than each could carry individually. All nest members – workers, larvae, queen, and
soldiers – share food, but not everyone eats for herself directly. Recently ingested food is available to those who do
not eat immediately. Each ant stores partly digested food
within a crop in its abdomen and freely regurgitates and
shares it with any other member of the colony. Ants meet-
ing in the bivouac or on the trails routinely stop and share
food, mouth to mouth, and thereby maintain a fairly even
distribution of what is available. Each ant asks for food
when it is hungry, and donates when it is full. By constantly
checking with other ants, it has direct knowledge of how
hungry the colony as a whole may be. These checks also
serve to maintain security within the colony by keeping out
ants of other species, or other colonies of the same species.
If an ant does not carry the scent of the colony it will be
driven off or killed.

Toward afternoon or early evening, as booty-laden
traffic from the raid continues, stirrings begin to develop

within the nest. A contingent of workers exits the bivouac and begins milling around the entrance, ants running up and down one trail and then another. Somehow, one trail is chosen over the others. Soon, workers carrying larvae and booty emerge and begin marching along the chosen trail for the evening's emigration. As the march develops, thousands upon thousands of additional larva-laden and food-laden workers stream out of the mass and follow the crowd. The bivouac is in full disintegration. The queen, surrounded by soldiers and her special guard unit, emerges toward the end of the procession. But she makes her way down the trail under her own power. There is always great excitement among workers as she passes with her entourage. Though considerably larger than any other member of the colony, she does not lay eggs during the nomadic phase and her abdomen has contracted to a much smaller size. She must walk the 100 or so meters to the next bivouac site, but compared to what she goes through in the statary phase, this is a rest period. The workers and soldiers will get a rest during that phase, while she is back at work laying 30,000 or so eggs each day.

There is often a great deal of booty to be carried from one bivouac site to another, and the 300,000 or so larvae that have to be carried grow to be as large as the ants carrying them. Each day's emigration is a strain on the colony's energy resources. To reduce this, workers ahead of the brood carriers improve the right-of-way by pushing aside sticks and loose earth, and smoothing out rough surfaces by laying their bodies down in hollow spaces so that others may walk over them. Chains of ant bodies form bridges over streams and crevices as workers link themselves together, leg to claw, and allow brood carriers and the queen to pass. After she is gone and traffic eases, ants loosen their grip on each other, bridges break up, chains fall to the ground, and the former bridging crew follows the rest of the colony along the trail to the new nest site.

Traffic jambs are routine at trail branches. Even before the brood procession arrives, incoming ants along one branch of the trail often take a wrong turn and collide with those on the other branch. Outgoing ants do not always know which trail to take either, often wandering down one before turning back and taking the other. As the brood carriers approach, traffic at the branch thickens further, with workers scurrying this way and that as if testing the waters to see which way to go. Some stop and explore nesting sites alongside the column and begin forming chains between tree root buttresses or beneath logs. Larger workers cling to overhanging surfaces, first embedding their powerful rear claws and then hanging, head down, until other ants crawl down and hook to them. But the colony usually passes through the first few trail choices before settling for the night, and these chains will begin to break up after the column passes. As the day darkens, the need for a nesting site becomes more urgent. Brood and booty carriers select a bivouac site, usually at another trail branch, by swarming around one or another chain of ant bodies as they form. This encourages workers to concentrate efforts at this site and to abandon all other locations. By the time the queen arrives the choice is clear and most of the brood has already entered.

The new bivouac is a half to a full meter in height and width, and consists of up to a million ants. Hundreds of ant chains dangling from an overhanging log or branch are linked together and tightened by cross-linking ants, so that the nest becomes smaller and tighter at the bottom, the better to shed rainwater. As each ant hooks its rear tarsal claws to the front legs of the ant above it, it stretches its body out and seems to go dormant, rarely moving until the next day. But activity in the bivouac continues until well into the night. Smaller brood-tending ants move larvae and food about through tunnels and passages in the bivouac, feeding the brood, tending the queen, and cleaning the nest. The queen and her guard settle toward the top of the nest with the brood just below. The size, shape, and internal structure of the bivouac adjust through the night and morning to changing conditions of light, rain, humidity, and temperature. As the outside temperature cools, the nest usually contracts slightly to keep the brood warm.

Toward the end of the nomadic phase, larvae spin cocoons, become pupae, and stop feeding. They are too difficult to carry about, and the colony must settle down for a few weeks. The statary phase begins at this point. The colony no longer requires the food supply to justify massive swarm raids on a daily basis, and no longer needs to emigrate each evening to find new hunting grounds. Smaller, less energetic raids continue on some days, but the colony can afford to settle into a semi-permanent bivouac in a more sheltered site. All other types of ants do not emigrate and do not need to synchronize their broods, so eggs, larvae, and pupae are found at any time in any stage of development. But army ants synchronize brood development to keep the colony as a whole ready to begin the next phase when the time comes. All eggs become larvae as the nomadic phase begins, and all larvae become pupae as the statary phase begins.

After a few days rest the *Eciton burchelli* queen begins laying eggs deep within the bivouac. Larger workers form the outer, structural layers of the bivouac while smaller workers scurry about within, moving, feeding, cleaning, and tending the newborn. The first laid eggs are given more food and care, develop longer, and eventually become the larger workers of the next generation. As the statary phase continues, new eggs grow into larvae and the previous brood (now pupae) is slowly metamorphosing into adults. When the adults emerge from their cocoons, the statary phase ends and the nomadic phase begins again. The cycle repeats eight to ten times each year, with each brood contributing up to 300,000 new adults to the colony, replacing those lost in the raids. Each worker lives no more than a few months (about three brood cycles). The queen lives on, cycle after cycle, year after year, reproducing workers and soldiers every brood cycle. The colony reproduces itself *as a whole* every year or so.

The colony, and not the individual, is the unit of evolutionary selection. Workers are females grown from fertilized eggs but are infertile themselves. With no young of their own, they give their lives to the success of the colony and assist the queen in her reproductive efforts. This phenomenon has puzzled evolutionary biologists for a long time: why would an organism work so hard to secure the survival of other organisms and for the success of another organism's genes, while giving up the chance to reproduce her own genes? This seems contrary to the principle of survival of the fittest, or at least of the hardest working. Each worker is a daughter of the queen, and thus shares one half of her genes with her, but this degree of relation is not enough to prompt the degree of altruistic behavior seen in insect colonies. William D. Hamilton, a British entomologist, pointed out in 1963[11] that hymenopteran workers (ants, bees, and wasps) are all sisters, but are much more closely related to each other than other animal sisters because they all get the same batch of genes from their father. This is because, unlike other animals, hymenopteran males develop from *unfertilized* eggs. Males have no father. They have only one set of genes, *all of which* they get from their mother. When a male ant mates with a female the offspring are more closely related than in most animals because their genes are less randomly mixed. Each daughter gets a random mix from her mother but an identical mix from her father. Where most siblings share half of their genes, hymenopteran siblings share three quarters. According to the principles of genetics, this gives each worker a greater interest in the survival of the colony because the colony as whole will pass on a large portion of her genes even if she has no offspring of her own. This explains the altruistic behavior of army ants and other hymenopterans from the standpoint of genetic survival. It does not, of course, explain the mechanism through which genetic similarity translates to coordinated behavior, but it explains insect social behavior in the same terms that we use to explain other types of animal behavior.

Despite the rigors of daily emigration in the nomadic phase and the mothering of several hundred thousand young during the statary phase, the army ant queen lives for five or more years. Most of her duties through the year are limited to reproducing individuals within the colony, but she is also charged each year or two with reproducing the colony as a whole. During the dry season she and the colony she rules raise a special "sexual" brood. Five or six fertilized eggs are laid and, with special nutritional attention denied to worker eggs, develop into fertile females, that is, into new queens. Two or three thousand other eggs are left unfertilized and become males. As the sexual brood nears full development the bivouac divides in two: the existing queen and those loyal to her on one side and the new queens and workers loyal to them on the other. Not all individuals develop loyalties at this point, but when the nest divides in the next few days, each will have to choose one queen or another. Chosen queens will survive and found new colonies, those not chosen will be abandoned and die in a few days. The existing queen is usually the first to emerge from the bivouac and, if she is not too old, has the best chance of being chosen. With her special guard unit still intact, she marches out of the bivouac with as many workers as will follow her. Next, one of the new queens emerges, and then another. The first is usually most successful in attracting nestmates, but if she is "unpopular," workers will abandon her and join up with one of the others. If two new queens attract enough followers, the existing queen will be left behind. Meanwhile, the entire bivouac is disintegrating and hundreds of thousands of workers are milling about in every direction, deciding which way to go. The chosen queens march off in opposite directions and the whole colony divides like an ameba, about half going each way. The next day, a small two-way connecting trail remains but soon disappears, and the separation is complete.

Meanwhile, the males have grown wings and are preparing to fly off and find new queens in other new colonies. But they do not leave yet. They divide themselves up between the two chosen queens, following one or the other to a new nest. They do not remain for long. As their wings mature, they try to fly, but are restrained by workers who seem to sense that it is yet too early. When the time is right, and other colonies have queens ready for mating, the males are allowed to fly off into the night, gaining as much physical and genetic distance from their home colony as possible. When they land, their wings degenerate and they begin sniffing and crawling about, seeking out trails of some new colony. But they must allow time for the scent of the home colony to wear off in order to gain the scent, and thus the acceptance, of the new colony. Workers in each new colony, having chosen their new queen, proceed to choose a male to mate with her. Many thousands are available, and most must be avoided, ignored, sealed off from the bivouac, or killed. The new queen mates only once in her life with only one of them, keeping millions of his sperm alive in her abdomen for as many years as she remains mistress of the nest. He, on the other hand, having spent his biological wad in one shot, will die the next day.

The question that always arises among human observers of army ant behavior is how does each ant know what to do? Who determines when to begin a brood, when to pivot the swarm to the left or to the right, how to construct the bivouac, and which trail to take? Is there a plan? If so, who makes it and how is it communicated to the field?

Much of the coordinated behavior of ant societies can be explained in terms of pheromones. Chemical signals determine which way ants go and what general behaviors they display upon arrival. There are distinct chemical signals indicating alarm, attack, food sources, recognition of nestmates, and many others situations. When food is discovered, enemies are encountered, or an ant is attacked or killed, chemicals are released by ants on the scene that bring appropriate attention to that location. But pheromones, even if they tell individual ants what to expect and where to expect it, do not tell them how to coordinate activities once they arrive. They cannot tell one worker to begin forming a chain at a particular location, another to begin somewhere else, and others to fill in the gap between. They cannot tell five or six workers to lift separate sections of a large piece of captured booty and carry it back to the nest. So how do they know? Army ants are almost completely blind and cannot possibly know what is happening a short distance away. No one of them "oversees" trail development, raid strategy, or bivouac architecture. Among human individuals these types of activities must be preconceived and orchestrated.

Close observation shows that these actions, though they follow precise patterns, are accomplished entirely through trial and error. Individual ants scurry this way and that, pick up pieces of food or brood, carry them about, then set them down and go somewhere else. Other ants later pick up the same items and place them somewhere else. Nobody seems to know what to do. Often one team of ants is busy undoing what another team is doing. When a nest is disturbed, for instance, some ants pick up larvae and carry them outside, while others take the same larvae back in. Others simply run back and forth, never accomplishing anything. But when a few ants stumble into successful behavior, other ants quickly abandon what they are doing and join in. If two or three ants are struggling to lift a large piece of spider meat and manage to move it, others stop whatever chaotic behavior they are in the midst of, and lend a hand. When the general flow of eggs or larvae is out of the nest, ants will stop carrying them back in. As soon as the queen enters a bivouac under construction, work ceases at all other sites. Workers disassemble chains they have begun and help complete work at the chosen site. This phenomenon is sometimes referred to as *stigmergy*,[12] or the stimulation of new work by successful work already accomplished.

Stigmergy is no more than trial and error, but it requires *recognition* of orderly patterns of work. This seems to be inborn. All animals (and perhaps all living things) have this capacity; it is only in the social species that it has developed into collective behavior. In some species of leaf-cutter ants, for instance, elaborate earthen nests complete with brood rooms, fungus gardens, ventilation shafts, and thousands of openings to the surface, are constructed over the lifetime of many generations of workers without any overall plan or supervision. Work begins with a period of chaotic back and forth motions and only takes shape when two or three pellets of earth happen to end up in a pile. Others notice the pile and begin adding to it. Order is perceived, not conceived.

It is only among higher animals that sufficient nervous capacity is developed to consider patterns of work behavior before they are executed. But where did the ability to think come from? Chaotic behavior patterns of worker ants trying to decide which way to turn or where to settle for the night simulate, in many ways, chaotic thought patterns as a problem is considered. We tend to think one way and then another before coming up with a course of action. Stigmergy, the development of orderly ant activity flowing in one direction and then another, simulates in many ways the coalescence of thought into decision. A *conceptual* realm of consciousness, consisting of numerous untried practical options, most likely evolved in higher animals from the recognition of orderly patterns in the perceptual realms. It became highly adaptive by obviating expenditures of enormous amounts of energy trying out numerous options physically before settling on one of them. "Looking before leaping," or envisioning ways to do things without physically trying each possibility, is an exceedingly efficient way to create order. Humans, on the individual level, have used this realm to great advantage. On the collective level, however, we continue to rely for the most part on stigmergy. We expand cities, synthesize new products, or march off to war without knowing what we are doing or why, hoping that order will emerge from the consequences. Were we better able to conceive and evaluate the consequences of collective actions before taking them we would reduce the "error" side of the process of human social evolution.

The other question inevitably raised in the study of insect societies is that of the "superorganism." Are we looking at so many individual army ants with interesting collective behavior patterns, or at a single integrated organism composed of many individuals? The question is, of course, the same as encountered in our study of sponges, only then it was between cell and organism and now it is between organism and society. As then, there can be no definitive answer and it is the question itself that most interests us.

Again, we are looking for a distinct unit of consciousness, a living *thing* that is the absolute thing and not just part of something else. We want consciousness to be inside of a discrete package to which we can relate from the standpoint of "our own" discretely packaged consciousness, even if the package has to be some sort of superorganism.

William M. Wheeler, a prominent early twentieth century entomologist, believed that "(The ant) colony is an organism, and not merely the analogue of the person."[13] He claimed, as we might, that for ant colonies and simple multicellular animals "...the same very general laws must be involved.... The colony is not equivalent to the sum of its individuals but represents a different and at present inexplicable 'emergent level.'" He noted that in ant colonies, as in many-celled animals, there is a clear differentiation between "germ plasm" (queen and males) and "soma" (workers). Maurice Maeterlinck in his *Life of the Bee* (1901)[14] spoke of a mystical "spirit of the hive" governing the behavior of individual bees. Charles Darwin was equally impressed by ant colonies. He did not initially understand the social dynamic of ant colonies and considered the existence of a non-breeding population (workers) a serious challenge to his theory of natural selection. It was only when he realized that the colony as a whole, and not the individual ant, was the unit of selection that he was able to accept his own thinking on evolution. But the reductionist tendencies of modern scientific thought have seriously damaged the concept of the superorganism. Edward O. Wilson, a prominent contemporary entomologist and specialist in ants, claims that, "Seldom has so ambitious a scientific concept been so quickly and almost totally discarded."[15] He claims that social behavior can be, or some day will be, fully explained in terms of artificially induced responses under laboratory conditions of isolated members of the colony. There is no forest, only trees.

A question more important than the existence or non-existence of the superorganism is how does the holistic dimension of insect society compare and contrast with the emergent holism of human society? The contrasts are the most glaring. Ant societies evolved around 100 million years ago, and if still evolving at all today, are doing it so slowly as to be unnoticeable. Human civilization (cities, writing, division of labor) is about 5500 years old and evolving at an accelerating pace. Ants (particularly army ants) are blind, nearly deaf, and small in body and mind. Humans are large, highly developed perceptually, and the most mentally complex of animal species. Division of labor is limited in ants to a dozen or so specific tasks (foraging, soldiering, brood tending, egg laying, waste removal, etc.) with about a half dozen body types (queen, males, soldiers, major, median, and minor workers, etc.). In human societies division of labor is many times more advanced (butchers, bakers, and candlestick makers), with only two basic body types. Army ants share food and other resources on a more-or-less equal basis. Human societies are hierarchical and almost never egalitarian. Ants are always altruistic toward their particular colonies. Humans display a broad range of tendencies in this regard. They are usually altruistic within their own society, sometimes toward societies other than their own, and at times to the species as a whole. Ant communication is limited to chemical symbols, which, though sophisticated by human olfactory standards, cannot carry great amounts of high quality information. Humans use auditory symbols, written symbols, and electronic images to communicate, all of which are virtually unlimited in information carrying capacity. Ant colonies may be said to be "aware" as a whole of danger, food supply, weather, brood condition, etc., but they have no specialized sensory organs or organisms. Human societies are aware as a whole of danger, food supply, weather, financial markets, etc. through technological instrumentation. Information is gathered by specialized individuals and mediated to the rest of society. Army ant bivouacs provide shelter, protection, and a degree of homeostasis (constancy of temperature, humidity, light, etc.), but most individuals remain in direct contact with the atmospheric environment. Human habitations provide an increasingly greater degree of homeostasis and separation from atmospheric conditions. Army ant colonies are composed entirely of ant bodies. Human civilizations incorporate inorganic technologies within the life of the community. Ant societies have specialized individuals responsible for reproduction of the society as a whole. Humans are reproductively unspecialized, and with the exception of some overseas colonies specific to certain mother countries, do not generally reproduce societies as a whole.

If the Earth were as large proportionately to human societies as it is to ant societies, the human superorganism would evolve in an entirely different manner. Separate societies under differing local conditions would adapt for millennia without commercial or cultural contact. Isolation would create genetic distinctions in each society and specific adaptive traits linked to ongoing technological evolution. Something in the body or mind, for instance, would be specific to a particular tool or software program. Such techno-genetic traits would be kept distinct from other competing societies and passed directly on to offspring societies. Human societies would begin to compete by reproducing themselves on the social level, as do ant colonies, and there would be great adaptive pressure to leave reproduction to specialized individuals. As it is, the Earth is not large enough to maintain this degree of isolation. Factors driving adaptation – climate, environmental change, resources, etc. – are global in scale and not local. Societies compete commercially and technologically, but they are not able to link genetic with technological evolution. Technology passes too quickly through the social membrane to the society next door, and adaptive advantage soon spreads evenly across the planet. We are not different enough to evolve separately and there is no competitive advantage to specialized reproduction.

With the exception of reproductive specialization, altruism and direct sharing of resources, human societies are generally more holistic than ant societies. This is difficult for humans to see, as they (we) are both object and subject. We are too close to what we are looking at and more likely to see details than overall patterns. Scientific methodology, which we apply to ourselves in such analyses, is generally reductionist in spirit, and more likely to perceive spaces between parts than wholeness. Scientists generally explain a forest as an interaction among trees. Viewed from a distance, and perhaps by species other than ourselves, the question of the superorganism, with or without its answers, is more appropriately applied to humanity than to army ants.

We do not share a phylum with army ants, and perhaps they and other hymenopterans are all that can be expected of social development among arthropods. But we do share a kingdom with them, and more fundamentally, a life force that is always trying to do new things. That life force is now doing something very new and different in our own phylum, and particularly, our own species. Its chances of adapting to an increasingly limited environment depend on how well it utilizes conceptual realms of consciousness.

IV
_____

The Structure
of Consciousness

How does consciousness get from one cell to another to form a sponge or a hydra, and from one organism to another to form an ant colony or the Roman Empire?

How does consciousness enter life? At what point do cells and cell colonies become conscious? Are ant communities and human societies conscious? Where is consciousness: in a cell, group of cells, or somewhere in between them? Where is social consciousness: in an individual, a group of individuals, or somewhere else?

Consciousness is whole, composite, and mysterious. Whole, in that it cannot be divided; composite, in that it has interrelated parts; and mysterious, in that it will not be understood. Consciousness is everything. Understanding is the relation of one thing to another; you cannot relate everything to something.

Wholeness is the enclosed cell, the organism, and the society. Parts are organic proteins, body cells, and individuals. The *functioning* of parts may be understood scientifically – proteins, cells, and individuals may be understood in terms of interactions – but the *experience* of parts, and of their composite whole, may not. Conscious experience is outside of the box and beyond the bounds of science. The structure of consciousness, or the integration of chemical, cellular, and individual experience into cellular, multi-cellular, and social experience, remains mysterious. It cannot be objectively analyzed. Yet it is immediately visible. You can see it, now: Your experience is the composite experience of your cells. What you see around you is the collective experience of society. It is as real as reality itself.

To the extent that consciousness shadows organic function, or is shadowed by it, the structure of consciousness may be inferred from science. Science tells us how cells interact physiologically to create an organism and how cellular experience is structured into the experience of an organism. Some cells absorb nutrients, others provide protection; still others send and receive electrical impulses. The wholeness of the organism is manifest in the ability of its cells to act together in an orderly manner: A cell or group of cells encounters a food source and relays an encoded message to other cells that alert still other cells to contract in a particular order. If the right contractions are accomplished at the right time, the food is captured, digested, and assimilated, and all cells benefit. Coordinated action indicates wholeness of being, or experience shared by cells in the organism.

Individual cells experience the stimulus, the message, and the contraction, but it is the animal as a whole that moves a leg or a tentacle. Shadowing the functional coordination of its parts is the *being* that is the animal.

That there is consciousness in an animal able to move a leg or a tentacle can be neither proven nor disproven. Neither can it be proven in a human being. We assume, for everyday purposes, that people we know and love and walk by on the street are experiencing something like what we experience ourselves, but there is no way to know this. I will leave this assumption unquestioned for the moment, noting only the great injustice of assuming that people are conscious and cats, dogs, monkeys, frogs, starfish, and amoebas are not. If there is a line to be drawn between human and amoeba consciousness, where is it drawn: Around people? Around animals? Is there no life in the eyes of the family dog? Frogs, ladybugs, cuttlefish, and planaria have eyes also. Plant cells have chloroplasts that react to light. We are most aware of life closest to our own kind, but it is grossly chauvinistic to assume that only humans and a few chosen servants are truly alive. There is no way to prove scientifically that *anything* is alive except in terms of the coordinated reaction to stimuli, which is shared by all living beings. We may refer to some life as *intelligent* and some not, but there is no clear line here either. Is what we call *intelligence* mere similarity to us? That there is no place to draw a line between conscious and not conscious means that there is no such line.

The problem of determining what is conscious and what is not stems from the problem of thinking of consciousness *in* an organism. The scientific tradition thinks of life in functional terms and tries to put life inside of function. The experience of being a cell is thought to be somewhere in the cell, or the being of an animal somewhere inside its central nervous system. But being is not physiological process; it is *looking at* physiological process. It is awareness of function, not function itself. This does not mean function exists only when we are aware of it; this means that consciousness is the experience of *seeing* and *understanding* physiological function (or anything else) and not a product of physiological function.

For the simplest animals, the experience of one of its cells is similar to that of the organism as a whole. Sponges have no specialized sensory or nerve cells. Their cellular experience is not encoded and delivered to a distance. Adjoining cells find out what is going on by "word of mouth" (which may be a bad metaphor for a sponge), or through contact with roaming amebocytes, the amoeba-like cells that move about between inner and outer cell layers. Cells that regulate the size of the osculum (bowl opening) probably sense and react to changes in water currents (or waste contents of the water), but they do not "hear about" what is going on in the rest of the sponge through any sort of internal medium. They hear about it through the medium of the water passing from one part of the sponge to another. The experience of the sponge as a whole is not, therefore, qualitatively distinct from that of its individual cells.

Hydras, unlike sponges, have nerve nets, but no nerve centers. The net provides a medium through which impulses move from cell to cell, but there is no processing of information. Everybody hears the same unedited message. It is as if each cell has its own telephone connecting it to every other cell, but nobody is in charge. Hydra cells differ in function and in how they react to what is happening, but their experience through the neural medium differs very little. If a cell has muscle fiber, it contracts in reaction to neural stimuli; if it has enzymes, it secretes. Large numbers of cells hear the same news at the same time, and coordinated contraction becomes possible. A signal that cells lining the gastrovascular cavity are getting hungry causes cells in the outer layer to change their shape at the same time, "somersaulting" the body to a new location. The animal knows that the food supply has dwindled and that the time has come to leave but, lacking higher sensory organs, it can have no idea where to go. All it can do is go elsewhere.

Every cell in the hydra is sensitive to external pressure, lack of food, etc., and responds to its environment in some particular manner – contracting, secreting, closing pores, etc. Cells experience nerve impulses, and there is an internal medium bringing the lives of individual cells together. A higher dimension of collective consciousness is created by the uniformity of life experience brought to each cell by the nerve net. This is what the hydra *is:* There is a quality to the organism's life that is not in the cells of which it consists. But the collective being of the hydra is limited by lack of specialized sensory organs. There is no information that some cells access and others do not, and therefore not much differentiation of experience among cells. The life of the hydra, though collectivized, is probably not, therefore, appreciably unlike that of its cells.

Central nervous systems found in higher animal phyla provide a greater uniformity of cellular experience that becomes like a screen or a blackboard for collective consciousness. Large numbers of cells in higher organisms experience the same stimuli at the same time. But unlike sponges and hydras, animals with sensory organs and central nervous systems experience a collective consciousness that is qualitatively distinct from that of the cells of which they are composed. Experience originates on the cellular level, but when encoded and processed by a central nervous system, it becomes sensory consciousness that no single cell is capable of experiencing. A retinal cell senses a photon, or a tympanic cell an air molecule, but neither *sees* nor *hears*. When a sensory cell feels the impact of a photon, it does not send the impact to the brain, it sends an electrical impulse to the brain. The impulse contains encoded information that is processed along with information from thousands of other sensory cells, and a *picture* is composed that no individual cell, retinal or otherwise, can *see*. Seeing is the being of the organism as a whole. The rise of collective consciousness among cells is associated with the functioning of this sort of intercellular medium created by the central nervous system.

Any information system, intercellular or interpersonal, requires a set of possibilities, only some of which are realized in a given signal. If there are no possibilities other than the signal, there is no signal. Information requires a *potential*, only part of which is actualized. An information system must consist of an array of possibilities, some of which happen and some of which do not. If the system is a telegraph, the potential is the possibility of a dot or a dash. If it is a blackboard, the potential is all the letters, numbers, and other symbols that *could be* written on it. If it is a television or computer, the potential is the screen with its thousands of tiny pixels that can be black or white, red, yellow or blue. The potential gives meaning to the actual. If there is no information potential, the random "click" of two metal plates that happen to come into contact has no meaning. It is just a click. But if the click is of a telegraph key it means that somebody is trying to tell you something. It could be this or that: a dot or a dash. Information potentials have the meaning they are given by their users, and have an artificial, "agreed on" quality. As a potential, a pixel screen carries much more information than a blackboard or a telegraph key because it delivers many more possibilities at a given moment.

Information potentials are the structural bases of perceptual consciousness in higher animals. Cells create media through which information is conveyed throughout the organism. There are both functional and experiential manifestations of these intercellular media. The neural connection of sensory to large numbers of other cells is the functional manifestation. A single sensory neuron may be *on* at one time and *off* at another, or a bundle of neurons may present a composite picture at a given moment, some neurons firing and others not. Their actual location in the potential creates a picture against a background. The more neurons connecting sensory cells to the brain, the more possibilities available, and the bigger and more detailed the picture. The experiential manifestations of intercellular media in higher animals are the potentials we experience in the form of space dimensions. Objects seen, heard, and smelled are actual perceptual information located at points in three intercoordinated information potentials.

Each intercellular medium is one of the five sensory realms. The three we have just mentioned, the olfactory, auditory, and visual, are *spatial* realms, in that they correspond to the three dimensions of space. The *temporal* realms are the chemical and tactile.[16] Sensory realms and their functional manifestations vary in size and importance among animals. Olfactory nerves and lobes are large in fish, reptiles, and most mammals, indicating a large realm of olfactory experience. They are small in birds and humans. Optical nerves and lobes are large in birds and higher mammals and smaller in reptiles, amphibians, and lower mammals. In speaking of the structural relation among the perceptual realms it is in *potential* consciousness that we are interested; the amount of *actual* information is less important. Humans, for instance, do not experience the rich olfactory world of the possum or the catfish, and do not have as highly evolved olfactory lobes; but we can smell. We are conscious in this realm. The conscious awareness of an organism is less a matter of actual than of potential experience; that it is alive and aware has to do less with what it smells, hears, or sees at a given moment than with the fact that it *can*, smell, hear, or see. Consciousness is not what is on the "screen"; it is the screen itself.

A single realm of perceptual consciousness is analogous to a telegraph key, a blackboard, or a computer screen, but the structure of an animal's perceptual consciousness as a whole is less complicated than three or four or five separate information systems. The animal could not function if its perceptual consciousness were a loose array of unrelated potentials. Separate systems for each sensory realm would result in the animal living in a separate "world" for each sensory realm. There would be no way to relate what is heard with what is smelled or seen. Perceptual consciousness has maintained its wholeness by evolving in an integrated manner where potentials are standardized and coordinated with each other. Each realm keeps its own potential, but potentials become structurally similar. Actual information remains distinct in each realm (there is no danger of confusing sound with light), but the potential for each realm becomes interchangeable with that of every other realm. The result is a single coordinated system where information in any of the five sensory realms is related directly to information in every other realm. The perceptual world becomes unified but multi-dimensional.

The ideal potential for an information system would be one in which there is room for all data but where important

data is highlighted. It would be both infinite and gradu-
ated, with the most important data up front. An ideal sensory potential would be one in which everything that is
actually heard, smelled, or seen, etc., has a place, but where
the most important sounds, smells, and sights are up close.
Less pressing information would have a place at a distance
proportional to its importance. In other words, it would be
a dimension. An ideal system for perceptual consciousness
as a whole would be one in which all five sensory potentials
were woven together into a single, integrated system radiating infinitely in all directions, where actual data in any one
realm would be potential data in every other. What an animal smelled could also be seen and heard. If potentials were
dimensions and the dimensions were inter-coordinated, the
animal would be able to see and touch what he hears, *and*
see and touch it *where and when* he hears it. It would be a
single, multi-dimensional world in which all sensory information had a place and a meaning. Only the potentials
would be coordinated; the animal would not *actually* see
what he hears if he is not looking at it, or hear what he sees,
if it is silent. But by knowing where and when information
is *potential* in all realms, he knows where to go to *actually*
see it, touch it, or taste it. This world is, of course, the real
world of space and time.

The structure of perceptual consciousness is, therefore,
the means by which the experience of individual cells becomes the being of a multicellular organism. Information
is encoded by sensory cells and delivered to other cells

through intercellular media manifested functionally by the organism's central nervous system. The medium brings a more-or-less identical set of stimuli to large numbers of cells, creating a uniformity of experience that becomes the experience of the organism as a whole. Messages from sensory cells travel in the form of electrical impulses to the brain, where their meaning is "read." To be informative, the message must be an actual within a potential; each "bit" of information must be only one of many possibilities. Each sensory realm has its own information potential to keep actual information from being confused with other realms. Potentials for all five sensory realms are standardized into dimensions and the dimensions coordinated into a single space-time world.[17]

In the case of vision, the structure of consciousness is traceable to the functional structure of the eye and optic nerve. The retina of higher animals has over a million receptor cells exposed to light, each connected to the brain through neurons of the optic nerve. Receptor cells come in two shapes: rods and cones. They are sensitive to a range of photon wavelengths from about 400 to 800 nanometers. Rods pick up dimmer light than cones, but do not distinguish between one wavelength and another. When they sense a photon they convert the "touch" they feel into an electrical impulse and send it to the brain. The brain "reads" the message as a shade of gray. Cones require more intense light, but are able to distinguish between wavelengths. Some pick up shorter wavelengths and encode what they "feel" in a message the brain reads as "blue." Others pick up the mid range and send a message that the brain reads as "green." A third group of cones senses only the "touch" of longer wavelength photons and sends only electrical impulses that are seen as "red." The manner in which the optic lobe of the brain functions to make sense of information from the retina is not as easy to discern. We know that the optic nerve consists of over a million neural fibers and that each fiber is connected to a network of associative neurons in the optic lobe of the cerebral cortex. An enormous data load is sifted and sorted to come up with the picture we call visual consciousness. To make the information meaningful, the brain presents it as a range of points within a dimensional potential. Actual physical objects are "seen" as ranges of points in space, each point being the "touch" experience of a receptor cell in the retina. As each of the million or so optic neurons contacting the brain is connected on the other end to a specific location in the retina, they must have a functional relation to the dimensional potential, but it is difficult to say what this relation is.

A sensory potential cannot be related in any absolute sense to a physical presence because it is not itself physical. Everyone experiences a sensory potential – you and I and the man on the street all have a "field of vision" within which we experience visual objects. But none of us can point to it, poke it, weight it or measure it – it is just there. There is no way to verify it. It is outside of the box and therefore beyond the bounds of science. But a problem arises when the visual potential is identified with space. It may look like space, in that you see objects arranged in spatial relations within visual consciousness, but we have been speaking here as if space were something that arises in the brain. If space is in the brain, what is the brain in? If space is a function of visual consciousness, and visual consciousness a function of the brain, how can the brain be inside of something that is inside of it?

This has to do with our assumption that conscious experience "shadows" physiological functioning. There is no causal relation between the two. We could as easily assume that physiological function shadows consciousness. The concept of a brain functioning in space is itself a conscious process, and therefore "located" somewhere "in the brain." This is a conceptual circularity that crops up when consciousness is forced into a conceptual category. There will never be a place for consciousness in relation to other things. There is doubtless a relation between function and the structure of consciousness, and I have referred to function in trying to understand consciousness, but this does not mean that consciousness is subject to the strictures of function. In fact, I believe quite the opposite. The physical world that we study through science, and within which we understand physiological function, *is* the structure of consciousness. Space and time do not exist absolutely or independently of consciousness, and consciousness is not *in* space and time. They are in it. What we call the physical world is a subset of reality and not reality itself. I believe this because of what science itself has discovered about space and time.

Let us return to the visual realm of perception. Light is best and most simply understood as visual consciousness itself. *Light is seeing.* Science tries to understand light as a physical phenomenon in space, that is, within an external structure of the universe. But if space is a structure of perceptual consciousness, as I suggest, light is not in space. Space is in light. This is demonstrated by the fact that "flaws in the fabric" of space-time discovered by science are flaws in the fabric of light. The field of vision experienced by multicellular animals consists of tiny points of light that are also the tactile experience of receptor cells in the retina. These points of light, or photons, have properties both of vision and of touch. Information in the visual realm is of wavelength, or color, while in the tactile realm the same information is of momentum. This is the well-known "dual nature" of light. Physical science has long known that photons act either as waves or particles, but has never known why. The phenomenon is explained by the fact that what the animal "sees" is what a cell "feels." The animal cannot see anything smaller than what the cell feels because its field of vision consists of an arrangement of points of tactile experience. Vision is too coarse to see extremely small objects because a single point cannot be an arrangement of points. The space-time structure of the visual "screen" is smooth and continuous on the everyday, *macroscopic* level, but breaks down on the cellular, or *quantum* level.

It is not visual experience alone that breaks down on the quantum level: Space itself breaks down. If space were a perfectly uniform, infinitely divisible structure of an independently existing universe we would be able to locate any physical object, no matter how small, at any point in space, no matter how small. But we cannot. Space-time is not perfectly uniform or infinitely divisible. Due to the physical limitations of space-time, we cannot know exactly where an extremely small object is if we know its momentum, or know its momentum if we know where it is. This is not because our measuring instruments are imperfect; it is because the dimensional structure *of the universe itself* is imperfect, and that imperfection is due to the structure of visual consciousness.

The quantum nature of light, the fact that it is not perfectly smooth and continuous, is the quantum nature of space and time. Space, time, and mass, distinct on the macroscopic level, are bundled together to form energy quanta on the subatomic level. This means that the multi-dimensional "screen" of physical reality breaks down for very small objects. If their momenta are so small as to become comparable to the momenta of the quanta that make up the screen, the screen can no longer locate them in space and time (and mass). It becomes too coarse to make simultaneous measurements of both momentum and location. What happens on the *visual* screen happens in *physical reality as a whole*.[18] This is the famous Heisenberg Uncertainty Principle. Aberrations in light are aberrations in space and time. The fact that the physical limitations of space and time are derived from the structural relation between the tactile and visual realms of perception is a strong indication that they are, in fact, sensory potentials.

Many other well-known physical phenomena show space and time to be within light and within the overall structure of consciousness. Science can find no physical medium in space for the transmission of light waves. This is most simply interpreted to mean that they are not in space. The dilations of space, time, and mass of relativity theory, the mysterious "role of the observer" in quantum mechanics, and the inertial properties of "matter"[19] are all subject to new interpretations when dimensions are understood as sensory potentials. Scientists define a line in space-time to be the "path" of light through space. They do not find light in space because they define space in terms of light. It is like trying to find a house in a room of the house you are looking for.

It should be noted also that our understanding of cellular physiology is, of course, from a multicellular point of view. Our use of space, momentum, force, and causality is how we, as multicellular organisms, have come to understand our own macroscopic level of being. It cannot be applied to understanding the "being" of cells that do not experience higher sensory realms of consciousness. We have to use these concepts, as they are the tools we have, but we should realize their limitations. In viewing what happens functionally within and between individual cells we are like physicists probing the quantum world with yardsticks and balance scales.

V

_____

Plumbing, Fortification, Telegraphy, and Reproduction

PLUMBING

Life flows in relation to water. It moves through water or
water through it, exchanging sustenance for waste. Life is
enhanced, water degraded. Life lives between the pure and
the voided, its health a measure of the difference. Its orientation is always upstream.

Plumbing happens when life becomes concentrated.
Inlets come too close to outlets; the flow of water slows,
the gradient narrows between pure and impure. There is
less difference between upstream and down, between good
water and bad. Life comes to a standstill or declines, await-
ing a more organized flow. Plumbing is organized water:
water in organs. It detaches the center of life from direct
contact with the stream; its inlet is farther upstream than

the center of life and its outlet farther down. Plumbing widens the gradient by channeling water, and providing for concentrated, orderly growth.

The first great plumbing revolution was a great success, though it concentrated on the outlet rather than the inlet. It improved the gradient by expelling processed water as far away as possible. As we have seen, the sponge cells that engineered this revolution concentrated themselves into a bowl-shaped organism that, by constricting its opening, thrust a stream of used water away from its inlets so as to avoid processing the same water again. Fresh, unused water naturally flowed into the bowl through small spaces between cells. There was no single inlet, so each cell remained in direct contact with the incoming stream. That this principle has succeeded is evidenced by the survival to the present day of many species of sponges; that it is an evolutionary dead end is evidenced by the fact that, as far as we know, more highly developed multi-cellular organisms did not evolve from this particular plan. Cells stepped back from the mouthless bowl idea and reorganized instead along entirely new plumbing principles. Hydras, jellyfish, and other cnidarians turned the bowl into a mouth, but used it for two-way traffic, as both inlet and outlet. The bowl evolved later into a one-way canal with separate inlet and outlet, becoming the fundamental plumbing principle for planaria, earthworms, butterflies, fish, and kangaroos.

Of animal societies, only humans have developed plumbing to any extent, and they only recently. Ancient cities without plumbing grew only to the extent that people could find clean water and sanitation on their own. Each remained in direct contact with water. This became increasingly difficult as populations became more concentrated. Inlets were too close to outlets. Diseases periodically spread throughout densely packed populations and life expectancy shortened as cities grew beyond their limit of sustainability. Birthrates remained high in the country and people migrated to urban centers despite health problems. Death rate in the cities almost always exceeded that in the country, but people kept moving to town. Throughout history many more people have come from the country to live in the city than from the city to live in the country. Cities then, as now, lived by in-migration and not by replacing their own populations. Once they reached a population limit based on water and sanitation (as well as defense, food supply, and other vital resources), urban populations stabilized or declined, even as in-migration continued. Attempts were made in many settlements to provide fresh water and drainage, with limited success. River or cistern water was often used for washing and bathing, while well and spring water was reserved for cooking and drinking. But always there was a limit to how much clean water could be gathered and distributed in densely populated areas. The ancient city that succeeded more than any other in bringing copious amounts of fresh, clean water to all its citizens was the city of Rome.

Rome was not the first to channel water or to build aqueducts. Other cities in Egypt and Mesopotamia and along the Indus and Yangtze valleys, some of them thousands of years older than Rome, had long since channeled river water for crop irrigation. Rome provided water less for crops than for people. It was the first to engineer a large, permanent, many-faceted plumbing system that supplied an abundance of fresh water to large numbers of its citizens, and to provide systematic sewage disposal as well. It was the first city to provide clean drinking water to a concentrated population of hundreds of thousands of people.

River water was not good enough. Situated as it is along the banks of the Tiber, Rome had ready access to all the river water it wanted. But the Tiber, like most large streams, was laden with silt, mud, and sewage – some of it natural, some man-made. To improve its position on the water gradient, Roman engineers looked beyond the waters flowing through the city to a source farther upstream – not on the Tiber itself, but on the Anio, a tributary originating in the Apennines to the east and south and circling the city to the north before emptying into the Tiber. Rome was extremely fortunate to have the Anio, and to have it nearby. It is smaller than the Tiber, closer to its spring-fed sources, and much cleaner. Of equal importance, the Anio is of higher elevation than the city; its waters could be brought in by the force of gravity. The farther up the watershed the Romans built their aqueducts, the cleaner the water and the more fall available for distribution within the city. Clean water flowed by a system of tunnels, above ground channels, and raised aqueducts, reservoirs, fountains, private homes and public baths, before emptying into the Tiber, all by a few hundred feet difference in elevation. The glory of Rome is in no small part the glory of the Anio watershed.

Vitruvius, a first century B.C. Roman architect and engineer, was highly aware of Rome's position on the watershed. But where we might think of a relation between water quality and the size of a watercourse, he thought of it in terms of exposure to the sun. "Water, however," he wrote," Is to be most sought in mountain and northern regions, because in these parts it is found of sweeter quality, more wholesome and abundant. For such places are turned away from the sun's course... But on the plains one cannot get supplies of water. And what there is, cannot be wholesome, because in the absence of shadow, the violent power of the sun catches and drains, by its heat, the moisture from the level field. And if any water is visible, the air calls out the lightest, thinnest and most subtly wholesome part and dissipates it towards the sky; but the heaviest, the harsh and unpleasant parts, are left in the field springs."[20] So it was into the hills the Romans went for their water supply.

The eleven aqueducts built in ancient Rome consist mostly of underground tunnels and stone channels just above ground level. Only relatively short portions are the well-known and widely celebrated raised archways. These are concentrated mostly in the final segment of each waterway to keep it at as high a level as possible as it approaches the city. Eight aqueducts pass through the Aurelian Wall near the Maggioria Gate at a high, level area known as Spes Vetas. (Only one on the east side of the Tiber, the

Aqua Virgo, does not.) From here they emptied into a series of reservoirs, fountains, faucets, and pools, sometimes mixing with one another, where they became available to Roman citizens, publicly for everyone and privately for those who could afford it.

Large stone and concrete settling tanks were built at the upper end of each aqueduct. Here, along the edge of the river or spring where water was collected, incoming currents slowed and sediments dropped to the bottom. Outlets were constructed on the far end near the top of the tank so that only the clearest water entered the aqueduct. These tanks were usually built in pairs so that water could be diverted into one while the other was drained and cleaned. These settling tanks were the extent of water "treatment" in ancient times.

Tunnels were by far the longest segment of every aqueduct. They were extremely laborious to build, but required less skilled workmanship and less maintenance than above ground conduits. Tunnels were also less vulnerable to enemy attack. They were built by laying out vertical shafts in a line of sight across a plain or hillside. When the shafts were dug to the appropriate depth they were connected with horizontal tunneling to the height and width designated for the aqueduct, usually about 3 or 4 feet across and 7 or 8 feet high. Tunneling is always slow work, as there is only room enough for one man at a time to chisel and dig at the bedrock, but with multiple vertical shafts, large numbers of workers could be kept busy in many locations. Workers would descend each day through the shafts and excavate in either direction until a connection was made with the next section. The shafts also allowed light and fresh air into the tunnel. After construction the shafts remained to facilitate access for maintenance, and to allow for the escape of air pockets that impeded the flow of water.

Of vital importance from beginning to end was proper elevation of the watercourse. Too little slope and water would not flow; too much and it would lose the elevation it needed to reach the city. In general, the steepest aqueducts sloped about 1 foot every 150 feet, the most level about 1 in 500. But keeping the slope within each tunnel was not so critical. As long as the outlet was appropriately lower than the inlet, a low spot would simply fill with water. Even a high spot inside the hill would not be a problem as long as it was lower than the inlet and no air was trapped in the channel. Outside the tunnel elevation became of critical importance. Conduits above ground were built to follow the contours of hillsides at just the right height to maintain flow. Raised conduits on archways were built only to cross between hills and to maintain elevation as the aqueduct approached the city. Arches were constructed of wedge-shaped "vousoir" stones with the water channel on top. Though difficult to cut and erect, vousoir stone arches are far more efficient in materials and labor than continuous stonewalls without arches.

It is hard for us to appreciate the difficulty of construct-ing and maintaining water conduits made of stone. The immediate practical problem is that stone is almost always flat or convex in shape, while the surface needed for conducting water is always concave. Stones at the bottom and sides of the channel had to be joined with water-tight mortaring that would withstand the stress of the spans, the weight of the water, the weight of the stones themselves, and the ravages of freezing and thawing. Any stress cracking due to temperature expansion and contraction or to differential settling of the substructure would lead immediately to leakage, and often to complete failure of the channel. That the Romans were able to overcome these difficulties is a testament to their engineering skills.

Channels were lined and waterproofed with *opus siginum*, hydraulic cement mixed with ground up pottery. In the absence of paper, plastic, glass, and cheap metal, nearly everything packaged in the ancient world was done so in ceramic containers of varying quality. Pottery was everywhere, whole and in pieces, and much of it recycled in the form of opus siginum. Old aqueducts were often re-lined with the same material. Cracks often sealed themselves after a time, as the high lime content of Anio River water left a build up that naturally filled small leaks between stones. But over decades Anio lime became a serious maintenance problem of its own, as it built up and filled in the main channels. Work crews were continually fixing leaks, replacing stonework, removing lime, digging out tunnel cave-ins, and rebuilding whole sections. Still, it is estimated that only about half of the water collected ever reached the city.

The architectural beauty and simplicity of the raised voussoir aqueduct is a lasting symbol of Roman civilization. The most magnificent examples are not in Rome itself, but built by Romans in Spain and France. The height limit of a stone arch is about 70 feet; where heights greater than this were required, a second or even a third tier of arches was added. The Pont du Gard near Nimes, France, a three-tiered aqueduct, is still standing at an impressive 180 feet. The two-tier aqueduct in Segovia, Spain is not only still standing; it is still in use.

It is ironic, however, that from a modern engineering point of view, the vast systems of above ground pillars, arcades, and stone-lined channels, as monumental as they may seem, are utterly unnecessary. Modern engineers are constantly pointing out that instead of the trouble of building raised arches to carry non-pressurized open water over low lying areas, the Romans should have used closed pipes that would distribute downhill pressure on one side of a low area to the uphill side. If the Romans had only known to enclose their water channels in pressure-tight conduits they could have run them at ground level or slightly below for their entire course, no matter what the intervening elevations might be. As long as the overall elevation decreases from inlet to outlet, water inside the conduit will flow naturally by seeking its own level downhill. Down hill pressure on one side of a valley will force water up the other side. But the Romans did not have cast iron or plastic and had no way to maintain water pressure in sealed conduits. With long sections of air-tight piping they would not have needed raised aqueducts to maintain a level water, and they could have laid all of their plumbing underground, as we do, leaving no architecture at all. The underground, pressurized conduit, in all its pedestrian utility, is the difference between ancient and modern plumbing.

There are a number of reasons why the Romans did not build pressurized conduits for their main water supplies. They understood the principles involved, but did not have the materials necessary to do the job. Most importantly, the materials they did use to build pressurized piping – lead and earthenware – were inadequate for large scale, high-pressure applications. These materials were practical only for low-pressure, low-flow applications. Lead piping was made from rectangular metal sheets ¼" thick and ten feet long. The long edges were folded together and overlapped, forming a seam that ran the length of the pipe. The seam was soldered with lead or tin, or simply sweated with enough heat to melt the join together. Sections of pipe were soldered together in butt joints or tapered on one end and flared on the other. Joints and seams were strong enough to carry water from a castellum to a house or a bath but never strong enough to withstand the pressure of millions of gallons of water rushing down a hillside. Lead was also expensive and dangerous to work with. Vitruvius was aware of the health hazards involved in lead piping and did not recommend its use. (Even though he described in detail how to make and use it.) Earthenware piping was equally unsatisfactory for high-pressure use. Earthen pipe sections were thicker and shorter, about 3-4 feet, with tapered ends joined with quick lime. Properly supported in critical areas, earthenware piping had some closed-pressure applications, but none on the scale of a major aqueduct. Open channel aqueducts were also easier to maintain. If a closed pipe became blocked there was no way to clear it. Sediment and mineral deposits continually built up in all water conduits, particularly at hard to reach elbows and corners. The best way to remove it was to send workmen directly to the spot. Inaccessible closed piping was doomed to eventual blockage and failure.

Open channel plumbing was the rule in the ancient world, and stone was its raw material. As inappropriate as it may seem to use large, disjointed, thousand-pound blocks of limestone to move water, that is what the Romans had, and that is what they made work. From large, bulky inert chunks of the Earth's crust they created the living tissue of Roman Civilization.

Rome became a great city and later a great Empire largely because it worked on its plumbing early on in its history. The system evolved for over 500 years, beginning well before Rome made its entrance onto the world stage. The first artificial watercourse was built in 312 BC, generations before the Punic Wars and centuries before the Caesars. It was built by a Roman censer named Appius Claudius (known also for having built the Appian Way, a major highway leading south from Rome). The Aqua Appia, as it was called, consisted entirely of tunnels and open channels, without the raised voussoir structures normally associated with the term "aqueduct." It consisted of stone-lined trenches stretching for about ten miles, following hillside contours where possible and boring straight through bedrock where necessary. Its source was a spring to the east of Rome, off a tributary of the Anio. It maintained a drop of 0.5%, a remarkable engineering feat for its time. Built at a lower level than later aqueducts, it entered the city underground in the region of Spes Vetus and served only the lower-lying areas of the city close to the Tiber. The Appia served for centuries and its water was considered one of the best in taste and healthfulness. It filled 20 castella, or reservoirs, below the Aventine Hill. It was later conducted on a bridge over the river to the Transtiber and served that region for a time, but lost importance as later aqueducts were built to serve more of the city, and perhaps fell into disuse during the height of the Empire. There is evidence, however, of its repair and reuse up to the 700's AD.

The Appia was followed by construction of the Anio Vetus in 272 BC and the Marcia in 144-140 BC. The Anio Vetus was four times the length of the Appia, reaching 43 miles up the Anio Valley, gathering water directly from the stream rather than from a spring feeding the river. Like the Appia, it was almost all underground, consisting of a tunnel 3.8 feet wide by 8 feet high. It included about 1100 feet of above ground structure, of which remnants of a cut stone bridge remain. It also entered the city underground near Spes Vetus but at a higher elevation than the Appia, with a terminus just inside the Servian Wall. Its higher elevation, along with its greater capacity, brought water to many more districts of the city. It served 35 castella, nearly twice that of Appia, and probably extended across the Tiber at some point. The Anio Vetus was the main water supply for the city for seventy years as Roman influence spread throughout Italy and the Mediterranean. Its water, however, drawn directly from the Anio, was often cloudy after storms, despite its settling tanks. When aqueducts with higher quality spring water were built in later years, Anio Vetus came to be used mostly for industry and irrigation.

The first true raised aqueduct, and some would say the greatest Roman aqueduct of all time, was built 130 years later. After the defeat of Hannibal and the Carthaginians in the Second Punic War, the Italian peninsula was firmly under Roman control and concern over the vulnerability of long sections of raised water channel was no longer a factor. The Aqua Marcia was longer (58 miles), of greater capacity, collected higher quality spring water higher up the Anio valley, and entered Rome at a higher elevation than either of the earlier waterways. It was mostly underground but included long stretches of voussoir arches spanning 5-6 meters each, with a rough-cut limestone cap covering the water channel. Like earlier aqueducts, it entered at Spes Vetus, but above ground, and was structurally incorporated into the Aurelian Wall, portions of which can still be seen. Larger than the previous aqueducts, it served a total of 51 castella throughout the city. Its waters were far superior to the Anio Vetus. Even after later conduits were built, it was widely known to have best quality drinking water in all of Rome. Praised through the centuries by poets and writers, it was considered special through the imperial period and beyond. Pliny the elder states that "The most celebrated water throughout the whole world, and the one to which our city gives the palm for coolness and salubrity, is that of the Marcian Spring, accorded to Rome among the other bounties of the gods."[21] Shakespeare, in Coriolanus, Act II, scene 3, has Brutus say that Marcia is "… our best water brought by conduits hither…." Shakespeare, always the poet and rarely the historian, does not mention that the events related in Coriolanus took place three and a half centuries before construction of the Aqua Marcia.

At the time of its construction, the Marcia was probably considered a final addition to the Roman water system. But the population continued to grow and demand for water increased. The Aqua Tepula was built only 19 years later. Shorter and of more limited capacity than Marcia, it was never more than a supplement to the existing water supply. Its source was among springs that form a tributary of the Anio in the Alban Hills to the southeast. Its main problem was not purity, but temperature. Our word "tepid" derives from the same root as the name "Tepula." It was never popular as a source of drinking water. Built several feet higher than Marica, it served new residential districts higher up the hillsides of the city. There were only 14 castella served by the Tepula. As it entered the city at Spes

Vetus its watercourse was constructed on top of arches of Marcia. (The Julia was later built on top of the Tepula.)

Remarkably little is known about the distribution of water once it reached the city. Rome has been occupied continuously for more than 2700 years. Streets, sewers, houses, squares, waterways, temples and public structures are constantly built and rebuilt over the centuries, leaving little trace of what came before. Much of what we do know is conjectured from small bits of direct evidence. The general picture is of each aqueduct terminating within the city in three types of main distribution tanks or *castella*. The first, with the highest priority, was for public use. A series of *salentes*, or waterspouts, issued streams of fresh spring water from the castellum that anyone could use to draw his or her daily needs. Below the salentes was a large *lacus*, or pool for dipping buckets. Water drained from the lacus into a sewage system, which often provided washing and latrine facilities on its way into the major sewers. The second and third castella were filled only with sufficient overflow from the first, so that the general public was always provided first priority. The second castella provided distribution to public baths and the third went to private customers. Through the wall of the third castellum each private customer was allowed to insert a *calyx*, or nozzle of bronze or lead about nine inches long and of standardized diameter. The amount of water he used was thereby regulated and paid for. From the calyx it flowed downhill through lead or earthenware pipes along streets, roofs, and byways, often underground, into the bathing room of his house. Some customers may have had valves to switch off the flow, but most likely it ran constantly, with perhaps some storage in tanks and other vessels. Toilets and washing facilities emptied into the drainage system downhill from the main castella during the imperial period there were a total of 247 smaller castella fed by secondary water conduits. These were scattered throughout all parts of the city. For supply in an emergency most of them had secondary supply lines from other aqueducts. If a castellulm's main aqueduct failed or was closed for repairs, the flow continued.

Well before the early aqueducts were built, Roman engineers were busy improving the city's drainage system. Legend has it that the earliest sewers date back to the period of the kings. Livy states that Tarquinius Priscus began the work that drained the Forum and the adjacent valleys into the Cloaca Maxima, and that his grandson, Tarquinius Superbus finished it, though this is disputed by modern scholarship. Livy reports that stonemasons did not appreciate having to work in the sewers. The main branch of the Cloaca Maxima, was constructed in some places of great blocks of Gabine stone, each about 1 meter by 1 meter by 3-4 meters, and in other places of concrete. The trench itself was about 3.2 meters wide and 4.2 high. It began near the Forum of Augustus and passed, with several tributaries westward under the Argiletum. A branch from the Velia passed in front of the Temple of Castor and then went on to drain the Lacus Inturnaw. The Cloaca Maxima then passed through the Velabrum to the Tiber. The mouth of the Claoca Maxima on the Tiber consists of three concentric rings of voussoir arches. A second system, the Circus Maximus had eight tributaries flowing into a single channel that emptied into the Tiber about 100 meters below the Cloaca Maxima.

Most of what we know about the workings of Roman aqueducts comes to us from an account known as "De Aquis Urbus Romae," (AD 97) written by Sextus Julius Frontinus (AD 40-103), a Roman soldier, engineer, and author. He was appointed curator aquarum, or water commissioner under the emperor Nerva. Upon assuming office as Rome's chief waterworks administrator, he began a comprehensive study of the major aqueducts, their histories and states of repair. It has survived to the present day as our best single document describing Roman water works as they themselves knew them. He begins his study with a statement characteristic both of his pride in the engineering accomplishments of his forebears and of the practical bent of the Roman mind: "With so many indispensable structures carrying so many aqueducts you may compare the idle pyramids or the other useless, although famous works of the Greeks."[22] He notes that the earlier waterways (the Apia, Anio Vetus, and the Marcia) were built at lower levels because the art of engineering level channels was not yet known or because they were purposely built underground to protect them from being cut off by enemies "since a good many wars were still being fought against the Italians."

Frontinus was concerned primarily with eliminating fraud and injustice in charging private customers for their use of public water. Many Romans had tapped the conduits illegally, or had increased their use without paying for it. Water use was measured not in volume or flow, but in terms of the cross section of a pipe inserted through the wall of a castellum, or public water reservoir. It did not matter how many gallons you got, or whether you turned off the water when not using it, all that mattered was how big the pipe was running to your house. This diameter was measured in turn by the width of a flat piece of lead folded into a pipe. A piece of lead five digits wide, for instance, could be folded into a q*uintaria,* or pipe a little over an inch in diameter, which became a standard unit of water delivery. Frontinus knew that a pipe angled upstream would deliver more water than one facing down, and perhaps that a pipe toward the bottom of a tank would deliver more pressure, but the concepts of velocity and pressure were ill-defined in the ancient world and he had no means to measure them in any case. Instead, he insisted that all calyxes, or water nozzles, be inserted into the castella at right angles.

The position that Frontinus assumed at the height of the imperial period, that of curator aquarum, became a permanent administrative post over a century before his time after a major overhaul of Rome's water system by Agrippa. Agrippa assumed an aedileship in 33 B.C., under Julius Caesar. The four existing aqueducts, the Appia, Anion Vetus, Marcia, and Tepula, were in a state of disrepair and no new aqueducts had been constructed for nearly ninety years. Roman armies had conquered Spain, Gaul, Britain, Macedonia, Asia Minor, North Africa and Syria, and the city had become de facto capital of the Mediterranean world. Powerful generals, with massive armies loyal to them personally had struggled against the Senatorial government and against each other for control of what had become a world state. The Empire had begun to assume its consummate form but the city had been neglected. Agrippa stepped in, under Caesar's appointment, and developed a comprehensive plan for water distribution and drainage throughout the city that became the foundation of Roman plumbing for the rest of its history as imperial capital. It is the workings of this plan, with some additions, that Frontinus describes in De Aquis Urbus Romae.

Agrippa re-engineered and rebuilt whole sections of the existing aqueducts, even the Appia, which by now was nearly 300 years old. He re-routed and rebuilt Tepula, mixing its waters with other, cooler waters to make them more drinkable. He then added two new aqueducts and improved the sewer system for the whole city. Finally he created a permanent commissioner and administrative staff to maintain the system. This became the curia aquarum, or water board, headed by Frontinus in 97A.D.

The first of the new aqueducts was Aqua Julia, named after Caesar himself. Its sources were a few miles past the source of Tepula in the Alban Hills. This was something of a shortcut. Four other major aqueducts, the Anio Vetus,

Marcia, and later the Claudia and Anio Novus, crossed the plains to Tivoli and then followed the winding main course of the Anio valley for the rest of their length, which was often twenty or thirty miles. The Julia and Tepula never entered the main valley of the Anio, but cut across the plain to the southeast, only about half the distance, and there drew their water from a tributary of the Anio. The Julia, with nearly double the capacity of the re-conditioned Tepula, had much of its raised portion built on top of the older water channel. Where they entered the city at Spes Vetus, they were incorporated into the Aurelian Wall above the even older Marcia. A section of this wall survives, with all three conduits still visible. Waters from the Julia were mixed with those of the Tepula to lower its temperature to a more drinkable range, but the combined waters were then re-separated upon distribution through the city. The Julia was always a very reliable source, even during drought. It supplied 17 castella in the eastern districts of the city.

Under Caesar, and later under Augustus, Agrippa developed a comprehensive sewer plan to drain every street in the city. He paved fountain areas and other public access points to insure proper drainage from terminal points of the aqueducts. He re-worked the sewers of the Cloaca Maxima and Circus Maximus and added a third sewer system at the Campus Martius, a low lying largely non-residential district in the northern part of the city where many new public buildings were under construction. He also built new embankments to regulate the flow of the Tiber within the city.

The Aqua Virgo was built to serve this part of the city. Begun 14 years after the Julia, it completed Agrippa's overall plan for distribution to the city. Where the Julia augmented water supply to the older eastern and southern districts, the Virgo supplied a large volume of water to the north and west, particularly Agrippa's public building projects in Campus Martius. It also continued across the Tiber on the Pons Agrippae (a bridge built by Agrippa) serving the northern Transtiber district. The Virgo is unique in being much lower than any earlier aqueduct except the Appia, and in not passing through the Spes Vetus, as had every aqueduct built before its time. Its source was a spring just upstream from that of the Appia, and its course followed that of the Appia for much of the way westward to Rome. But as it approached the built up portions of the city, the Virgo took a dramatic turn northward, skirting around the higher elevation of the Spes Vetus and then south again to the Campus Martius. It did not, therefore, interconnect at any point with other channels. It is also unique in that it is still functioning today, with a terminus at the Trevi Fountain, though the fountain itself is probably a later addition.

The first Roman aqueduct outside of the Anio River valley, the Aqua Alsietina, was built in 2 B.C. after the death of Agrippa. It was constructed under Augustus on the west bank of the Tiber. Originating at Lago di Martignano and Lago di Bracciano, far to the north, it was not spring fed and its waters were considered unhealthful.

It was not widely distributed (there were no castella) and not generally used for direct human consumption. Other aqueducts, including the Virgo, Marcia, Anio Vetus, and Appia already served low lying areas of the Transtiber for human consumption. It seems that the Alsietina was used primarily for irrigation and industrial purposes, including a fish hatchery. Unlike aqueducts that crossed the river, it entered the Transtiber at a high elevation, and water pressure produced by its descent into the city could be used to power industrial mills. But along with the low quality of its water, it produced a relatively low volume, and was probably abandoned in the next century after construction of the Aqua Traiana.

Agrippa's comprehensive system of sewers and waterways brought the level of public health within the city to unprecedented heights in the ancient world, but within fifty years of its completion it was showing signs of strain. Peace and stability in the early years of the Empire brought enormous new wealth to the city and large numbers of new people. The demand of common citizens for clean water and the demand of the wealthy for luxury living lead to the construction of two new major aqueducts in 52 A.D., the Aqua Claudia and Aqua Anio Novus. Built one on top of the other, the two snaked 43 miles up Anio valley, paralleling the older Marcia and Anio Vetus. (That this double line is considerably shorter than the Marcia is due to its more direct course.) The five miles closest to Rome were supported on voussoir arches. The bridge carrying both channels from

Capannell to their entrance into the city is considered the most impressive existing example of Roman aqueduct architecture. The Claudia's source is several springs feeding the Anio far up the valley, while the Anio Novus draws its water directly from the Anio even farther upstream. Their waters were kept separate the entire length of the double system, the Anio Novus in a channel on top of the Claudio. The Claudio's water quality was considered far superior, as unlike that of the Anio Novus, it was not muddied by storm water. The Claudia was built of masonry, the Anio Novus built on top out of concrete. The two water streams were partly mixed in a castellum at Spes Vetus before distribution, even though this mixture lowered the quality of the Claudia. (A portion of the Claudia was kept separate for use at the imperial complex on the Palatine and surrounding residential areas.) Together these two supplied far more water than any other aqueduct in the city: 92 castella. This was one third to one half of the total in Frontinus's time. They also entered the city at the highest elevation and could serve every district. Together with Agrippa's system this new system supplied the city's needs for centuries after its construction, despite continuing population growth.

Only two aqueducts were added after Frontinus, the Aqua Traiana in 109 AD and the Aqua Alexandriana in 226 A.D. The Aqua Traiana, on the west bank of the Tiber, was largely a replacement for the Alsietina. Its course followed that of the earlier aqueduct northward to the region northeast of the Lago di Bracciano, but it drew its water from springs near Trevignano. Its waters were, therefore, far superior. The Appia, Marcia, and Virgo already supplied the Transtiber over bridges, but their supply was of limited volume and, because bridges over the Tiber were not elevated, they could reach only low lying areas. The Transtiber was a growth area of the city the first and second centuries, and lacked the quality and distribution of water enjoyed by the older districts of the city. The Traiana provided high volume, high quality water at sufficient elevation to be available to the entire Transtiber. It also probably crossed the Tiber from west to east at some point, and served as an emergency supply for the entire city. The Traiana was the last of the great Roman aqueducts.

Aqua Alexandriana was build over a century later. Originating at springs of the Anio watershed to the east of the city, it was constructed at low elevation and its use was limited to supplying the baths built by Emperor Alexander Severus in the Campus Martius. It entered the city at ground level at Spes Vetus and did not connect with other aqueducts.

By the early second century A.D., the Roman plumbing system was complete. It remained in operation throughout the following centuries as the population of ancient Rome reached its climax, supplying the city with something in the range of 100 million gallons per day. Aqueducts, castella, settling tanks, and supply lines within the city were constantly repaired and rebuilt, but no major new water conduits were built, with the exception of the Alexandriana.

Every district of the city was served by three or four lines, and some of them by six or seven. When any one system broke down or was closed for repairs Roman citizens did not have far to go for an alternate source of water.

Historical evidence indicates that the system was still operating at the time of Constantine in the 4th century. In the 6th century, aqueduct channels were cut during the Gothic siege and then blocked by the Romans themselves to prevent their use by the enemy as a means of gaining access to the city. The city's population, by now much reduced, was forced to occupy lower lying areas in the Campus Martius. Much of the population remained there throughout the Medieval period, doing the best it could to find fresh water in cisterns and streams, and perhaps in the still functioning Virgo. There are records of brief periods of restoration of the Traiana, Marcia, Clauda, and Virgo in the 8th century, and of the Claudia in the 12th. The Virgo was restored several times in the 15th and 16th and 18th centuries, and flows through the city today.

Millions of years ago rainwater falling on the western slopes of the Apennines, gathered in the brooks and springs of central Italy, streaming down the winding course of the Anio Valley to the Tiber. Thousands of years ago people in small settlements along the valley collected water in gourds and buckets. Hundred of years ago vessels of rock and mortar sprouted and grew up along the Anio from the Tiber, rooting through hills and bedrock, under mountains and across the plains, immersing hollow masonry probes into the course of the stream. Fresh water from the stream flowed through concrete inlets and down stone-lined channels, around and through bodies of a concentrated human mass of houses and baths, temples and fountains, to the flower of the ancient world. Spent waters from streets and houses were gathered by gravity into stone and concrete ducts and then flushed to the river. Force became order. Gravity became life. The stream did not create Roman civilization, but neither did it merely make it possible.

FORTIFICATION

Fortification is to keep bad people out, in the aggregate. It is a wall of stone, earth, steel, or wood: circular, triangular, square, polygonal, or irregular, and may incorporate natural defensive features, including cliffs, rock faces, rivers, or ocean frontage. It may be a temporary refuge, a castle, a walled city, or an artillery battery. A temporary refuge provides safety during a raid, allowing villagers to live through a day or two of disturbance and return to their homes outside the wall. Raids destroy crops, burn buildings, and slaughter animals, but raiders soon run out of things to do and leave. A simple refuge cannot withstand a prolonged, formal siege. A castle or a walled city provides longer-term protection. There is always stored food and a well, cistern, stream, or other steady water source. People inside are safe for months, even years, and continue to live, work, metabolize and reproduce as best they can within the wall. Only a well-supplied siege army can starve out a fortified city or to breech its walls. An artillery fort, unlike a castle or walled city, defends an area larger than itself. It provides, in combination with others of its kind, a *defense in depth*, enclosing a large territorial expanse against foreign invasion. It often becomes a focal point of intense combat and is rarely used as a refuge.

There are three features of any fortification system: an outer wall, defensible gateways through the wall, and an interior stronghold somewhere within the wall. The wall may be a wooden fence, an earthen rampart, or a stone or concrete barricade. It must be strong enough to withstand whatever stones, clubs, darts, arrows, spears, bullets, cannonballs, or battering rams are brought against it, and high enough to keep invaders from climbing over. But strength and height alone do not make good fortification. They create a barrier to defenders as well as to invaders. A good fort design seeks not to avoid combat but to give advantage to defenders in the event of combat. The main wall or *rampart* protects defenders but also allows them to shoot over, around, or through it. It has a distinct directional gradient: missiles pass more easily out than in. One way to achieve this is to build a low wall or *parapet* on top of the rampart that defenders can fire over and hide behind between shots. The parapet is usually a relatively thin vertical extension of the much thicker rampart wall, creating in effect a single exterior face with a *wall walk* on top of the rampart and behind the parapet. Defenders can easily move to any point along the wall walk and shoot down on attacking forces. The height advantage makes it easy for defenders to see and gives them greater missile range. A common improvement on a simple breastwork is a parapet with alternating high and low sections. Defenders stand behind the higher sections, or *merlons*, between shots or while resting, and move to the lower sections, or *crenels*, only long enough to deliver a volley. (Crenellated parapets are also known as *battlements*.) *Arrow slits* are narrow openings in the wall that provide further protection for defending archers, but reduce visibility and accuracy. They are usually designed for close range combat, as when invaders are up against the rampart. *Gun loops* for cannon are designed for the same purpose, but larger and rounded, usually with a narrow upper extension for sighting. Surrounding the rampart there is usually a dry ditch or moat, and perhaps a secondary wall with its own ditch. Parapets, crenels, slits, loops, and moats all weaken the rampart to some extent, but often give the home team enough advantage to fend off much larger invading forces.

A common problem encountered in simple wall designs is that where defenders have great advantage at a short distance from the wall, they often lose advantage up against the wall itself. Besiegers who work their way through outer defenses can gain control of a short section of wall in order to bring up ladders, tunnelers and battering rams to penetrate it. When a long straight wall is attacked at any single place, defenders have the disadvantage of firing sideways

on attackers from most places on the wall. Rounded walls
are worse in that they reduce the point of defense to a small
area just above attackers. Corners in the wall provide blind
spots that attackers can use to advantage. To counter this
difficulty, towers and bastions are often extended out beyond the main wall at corners and at regular intervals along
straight walls. Towers are higher than the wall, and reach
far enough past it to allow archers or gunners to fire parallel to the wall at the flanks of attacking forces. The spacing
of towers is, therefore, never more than twice the effective
missile range. Very few fortifications are built without towers. *Bastions* serve similar purposes but are larger outer defense works at the same level or lower than the main wall.
Often projecting from corners of the main curtain wall,
they protect it with flanking fire and add forward-based
offensive capability. Bastions are often added to older fortifications with the advent of modern artillery because they
provide more area for large gun emplacements than towers,
and expose less vertical surface.

The second feature common to any fortification is the
defended gateway. A defensive wall cannot keep everybody
out all the time. It must be a semi-permeable membrane.
Most castles and fortified cities were rarely attacked, and
had to function on a daily basis without an excess of artificial impediment. To allow daily passage through the walls
without compromising security, large defensive structures
were built wherever gateways penetrated the curtain wall.
Often a gateway consisted of a chamber with inner and

outer gates extending for some distance into or out from the main wall. Towers were built at both ends and directly above the chamber. Along the sides were separate rooms with arrow slits or gun loops through the wall to the chamber. Attackers penetrating the outer gate found themselves trapped against the inner gate with fire coming from both sides and above. Outside the curtain wall retractable bridges spanned moats and dry ditches. Well-designed city gates were essential to any overall plan of fortification but their usefulness usually exceeded their defensive features. Attractive gateways and city walls were always a source of civic pride and served as a symbol of a city's community spirit. Heavily ornamented and architecturally intricate city gates greeted visitors with an imposing first impression. Tolls and tariffs were collected here from travelers and merchants upon arrival at the city. Gates were often closed at night to provide security from thieves and wild animals.

A third feature of nearly any system of fortification, beside the outer wall and gateways, is the interior stronghold, also known as a *keep*, a *donjon*, or a *citadel*. Its primary purpose is as a last defense against invaders who have breached the gates or the wall, but it always serves a number of secondary functions, including a watchtower, a command center, and residential quarters for the garrison or the lord of the castle. It is often a point from which all parts of the fortification can be seen, including the courtyard or town square from which a final assault would be launched. Food supplies and a water source are usually located here.

Examples of fortifications that include these features are Goltho, Kano, Biejing, Carcassonne, and Rhodes.

Goltho,[23] in Lincolnshire, England, was first built around 850 A.D. by the Anglo-Saxons as a defended enclosure. It was converted in 1080 to a small castle by the Norman conquerors of England under King William I. The enclosure is a small circle measuring only about 35 meters in diameter with a massive outer rampart and a small courtyard and fortified mound on the northeast side. It is surrounded by a moat with a bridge and evidence of a gatehouse. The site is nearly devoid of stonework, except for a basement or cistern on the mound that was built under the keep, or watchtower. In addition to the timbered keep, evidence remains a wooden palisade fence on the earthen rampart and a wooden residential hall that took up nearly half of the space within the enclosure. There is no well. In 1150 the upper portion of the mound was leveled, the courtyard was filled in with the spoil, and new, larger buildings were constructed, taking up nearly all of the space within the ramparts.

Kano[24] is a large city in northern Nigeria partly within a walled enclosure dating back a thousand years. The walls are earthen ramparts nine to twenty meters high, faced with mud brick. There are fifteen wooden gates at points throughout the circumference, each covered with sheet iron and flanked by guardhouses. There is a dry ditch outside and inside. There are remains of an older wall within. Construction phases occurred in the twelfth, fifteenth, and seventeenth centuries. The existing structure is 24 kilometers in circumference and built to secure an entire city of 30,000 to 60,000 people, complete with gardens, pastures, streams, and ponds. It was built to withstand a prolonged siege, but not to withstand modern artillery. A British colonial force easily took the city with a single cannon barrage that leveled a portion of the wall in 1902.

Beijing,[25] China, is enclosed by an enormous brick wall built mostly during the Ming Dynasty (1368-1644). The wall is 17 − 23 meters thick at its base and 11 − 13 meters high, making it considerably more massive than western fortifications. It has a wall walk, crenellated parapets, and square bastions at corners and wall angles. Towers and heavily timbered inner and outer doors protect the gateways. Marco Polo visited Beijing in 1280, before much of the existing structure was built, and remarked that the guardhouse at each gate housed a garrison of 1000 men. The entire fortress enclosed a total of 26 square miles and was surrounded by moat.

There are four parts to the city, each with its own wall. On the north side is the Tartar City with six gates to the exterior. At its center is another wall enclosing the Imperial City, with gardens, palaces, and lake. Within the Imperial city is a third wall and second moat entirely surrounding the Forbidden City. Within this inner sanctum is the palace of the emperor and other buildings. During the era of

Manchu rule, the Chinese were not allowed entrance to the Forbidden City. To the south of the Tartar City is the fourth part of Beijing, the Chinese City, built against the Tartar City with a similar though slightly lower wall with seven gates to the outside. The wall between the Tartar and Chinese cities has three gates, each with defensive towers.

At the west gate of the Tartar City is an enclosed gateway with inner and outer gates. One tower is directly above the inner gate and a second tower is on the middle of the outer wall. The outer gate, rather than built below this second tower, is on the north portion of the outer enclosure, at right angles to the inner gate. Invaders forcing their way through the outer gate would have to turn to the left to face the inner gate, exposing their more vulnerable right side to defensive fire from the second tower. (The right side is more vulnerable because shields are universally held by the left hand to leave the right hand free for swords and spears.)

Marco Polo writes that at the time of his visit the Chinese were unfamiliar with the battering rams and siege engines that had been in use in Europe and Near East for fifteen hundred years. During his visit to the city of Siang in Hupeh province, he found the city under siege by Kublai Khan. After three years of stalemate, the Mongols were preparing to lift the siege when the Europeans showed them how to build a projectile engine that could throw rocks of up to 300 pounds over the city walls. Massive boulders crashing through walls and into houses apparently demoralized the population to the point of surrender.

Carcassonne[26] is a fortified town in southern France 90 kilometers southeast of Toulouse. It is located in the gap between the Pyrenees and the Massif Central at the crossing of traffic routes leading from the Atlantic to the Mediterranean and from central France to Spain. There are two complete walls circling the city, with a *list*, or defensive corridor, between them. The inner wall originates in Roman and Visigothic times, rebuilt by the Trencaval noble family in the twelfth century. The town became part of France when King Louis IX (Saint Louis) captured it in the thirteen century. Louis and his son Philip III began construction of the outer wall in 1247. Both walls still stand, making Carcassonne the most intact of Medieval European fortified cities.

The walls enclose a small city, complete with a fortified chateau (Comtal) that served as a citadel, and the basilica of Saint Nazaire, built by the Trencaval family. The outer wall is encircled by a dry moat. Fifty-three towers are stationed along both walls, allowing height advantage and flanking fire on enemy forces up against the outer wall. In the event of a breach of the outer wall the towers also provided protection for defenders firing on enemy forces in the lists. On the north and south portions of the walls, two of these towers span the entire space between the walls, blocking enemy passage along the list. There are wall walks, crenellated parapets, and arrow slits in the walls at critical points. But there are no gun loops, as the entire fortification was built before the invention of gunpowder. The outer wall provided a second line of defense and effectively prevented the approach of siege engines that might threaten the inner wall. There are six *posterns*, or secret passageways through the inner wall and at least one through the outer wall. These were high enough above ground to require a ladder, and were probably used by messengers or suppliers during a siege, or perhaps as an escape route. There were only two gates through the walls, at the east and west sides, each heavily fortified with strong towers on either side. Rainwater was probably gathered from roofs and stored in cisterns, though there is a fortified passageway leading to a water source just outside the walls.

Carcassonne was considered impregnable at its time. During the Hundred Years War (1337 – 1453), the city was attacked by Edward, the Black Prince, son of King Edward III of England. The English destroyed the *Basse Ville*, or lower city built outside the walls, but were never able to penetrate the fortifications. Cannon fire was introduced to Europe for the first time a few years later in the same war, but the town managed to avoid fighting at that time. There is little possibility that its walls could have withstood an artillery barrage. The Treaty of the Pyrennes in 1659 transferred the province of Roussillon to France, moving the border farther south and west, all but eliminating Carcassonne's military significance. Had it remained a frontier town, its walls would not have survived the gunpowder age. The fortifications were abandoned and fell into disrepair. Under Napoleon the city was struck from the roster of official fortifications and the French government later issued a decree to have them demolished. An uproar ensued in the town and the mayor led a movement to restrore them for purely historical reasons. The restoration was accomplished with great care and skill, but without historical accuracy. Nineteenth century restoration standards were not what they are today, and the reconstruction is not altogether authentic. The steeply-pitched tower roofs so familiar in tourist guidebooks were originally of much lower slope, as is common in southern climates without snowfall.

Rhodes[27] is an island in the eastern Mediterranean with a fortress at its northern tip. Its walls were originally constructed by the Knights of St. John of Jerusalem in fourteenth century and later adapted to gunpowder age technology. Three outlying towers on narrow points of land protect the Ancient Harbor to the north of the city and the Grand Harbor to the east. Each tower had water and food supplies and was capable of independent defense. Ramparts at the main fortification are 10 meters high and 3 to 4 meters thick and reinforced to 13 meters thick in some places on the interior. Parapets are crenellated. A moat surrounds the entire wall, with outer ramparts and a double moat to the west and south. A section of wall to the southeast is without a second rampart or outer moat, and proved the most vulnerable to attack.

Towers were constructed at the corners and intermittently along the curtain wall. They were later lowered to the height of the curtain when adapted to artillery defense.

A citadel was built into the wall on the north side of the city facing open country. The wall between it and rest of the city had three gates but is now gone. The city was originally constructed with eight gates to the exterior but three were blocked to improve security. One of the gateways, the Marine Gate, opens directly onto the Grand Harbor with towers on either side.

The Ottoman Turks besieged the Fortress in 1480. Long expecting the attack, Pierre d'Aubusson, Grand Master of Knights of Rhodes, had strengthened the walls to withstand cannon fire and placed cannons behind the parapet along the wall walk. A chain boom was stretched across the entrance to Grand Harbor to prevent ships from entering. The Turkish fleet appeared on May 23 with as many as 100,000 soldiers and sailors and made a shore landing west of the fort despite strong resistance. An infantry charge on the west wall was initiated almost immediately, but repulsed with heavy losses. It was followed by a second assault and second repulse. The Turks then concentrated their attention on the tower of Saint Nicholas, one of the outlying defenses on a point of land north of the city protecting the Ancient Harbor. They began with an artillery barrage from the west side of the harbor, destroying the outer walls of the tower, then launched a combined land and sea attack, scaling the tower walls with ladders and grappling hooks. The Knights fired on the Turks from the citadel and released fire on the ships in the Turkish fleet. Seven hundred galleys were burned and sunk, and the Turks retreated.

Later, when d'Aubusson noticed the Ottomans preparing to attack the southeast wall, he deepened its moat and began reinforcing it from inside. The Turks opened a cannon bombardment and breeched the wall in several places, then built breastworks on the opposite side of moat to prepare for a final assault. Under the cover of darkness 50 knights slipped through a casement at the foot of wall, crossed the moat and used ladders to scale the far side of the ditch. Surprising the enemy, they destroyed the breastworks, forcing the Turks to abandon the southeast wall.

The attackers next resumed their assault on the Saint Nicholas tower, this time using a land bridge to cross the ancient harbor. The bridge was constructed on land and carried, under fire, by 30 boats across the harbor and set at the base of the tower. A second land and sea attack began, but the land bridge was hit by cannon fire from the citadel and destroyed.

The second attempt on the tower proving a failure, the Turks renewed their artillery bombardment on all sides of the land wall. They concentrated once again on the southeast wall, which the knights had partially repaired after the previous bombardment. The Turks built a new rampart on far side of the moat and brought their cannon into place. Under protective fire from the cannon, they began construction of a ramp across the moat and up to the curtain wall. But d'Aubusson had a large catapult brought into place and hurled huge stones on the work under progress, doing great damage. The Turks could not continue and were again repulsed. The artillery bombardment resumed from all sides and the Turks managed to bring down the walls in several places. Cannon fire over the walls and through breeches in the walls destroyed many buildings within the city. For a third time, they stormed the southeast wall, and this time scaled the ramparts and drove off the defenders. The Knights regrouped and pushed them back, but were soon themselves driven back a second time. After two hours of hand-to-hand combat the Knights prevailed and the Turks retreated. Defenders pursued them beyond the walls, and the retreat turned into a rout. The Grand Master was seriously wounded in the fighting, and was thought to be dying.

After three months of combat and 9000 Turks killed, the siege was finally broken. Despite extensive damage, defenses held at every location. D'Aubusson recovered from his wounds and began the long process of repairing the destruction and preparing for the inevitable return of Turkish forces. They did not attack again until 1522, forty-two years later.

In June of that year they arrived with 300 ships and as many as 120,000 troops. Defenders numbered about 3500. Fighting began with cannon bombardments and daily infantry assaults. Throughout the summer months the Turks struggled to reach the wall but were beaten back every time. Finally, in early September, tunnelers reached the north wall and exploded two gunpowder mines beneath it, filling the moat with debris. With the wall breached and the moat

filled, the Turks charged on foot. Three times they reached
the breach and three times were driven off by defenders.
An artillery barrage with infantry assaults on other parts of
the wall continued throughout the fall. By December both
sides were exhausted and demoralized. The Turks were
plagued with disease and high casualties, and the Knights
with dwindling supplies and lack of reinforcements. On
December 22 a truce was signed surrendering the city to
the Turks but allowing the knights safe passage. They were
provided with ships and sailed to Crete, later moving to
Sicily, and eventually, to Malta.

In time, gunpowder ruined fortification. Though ad-
aptations were made, it was cheaper, generally, to knock
walls down than to build them up. Stone, resilient to arrows and hand weapons, gave way in many cases to earthen
ramparts that could better absorb the shock of heavy iron
projectiles. Square towers that had for centuries provided
flanking fire along curtain walls became rounded to deflect
cannon balls. High vertical walls that protected defenders
from archers, cavalry, and barbarian hoards proved too big
a target for cannon fire and gave way to lower profile battlements. As vertical exposures were reduced, horizontal dimensions increased, absorbing huge amounts of space both
outside and inside curtain walls. Large pentagonal bastions
were built out past the wall, displacing suburban neigh-
borhoods. Mounds of rubble and excavated earth thick-
ened walls from within, reducing the usable space they defended. As urban populations burgeoned and cities spread

outside their walls, the prospect of building newer, longer, thicker, cannonball-proof walls became prohibitively expensive. Throughout the Middle Ages European cities had, like arthropods, shed their older walls and built new larger ones, but the molting process could not keep up with the growth. The prospect of defending an entire city was gradually abandoned. Fortifications became forts: smaller and stronger outlying structures built for military and not civilian occupation.

Forts themselves became obsolete as warfare became increasingly mobile. No matter how strong, well planned, and heavily-gunned a fort may be, an enemy must be convinced to attack it or it is useless, except as a deterrent. A mobile enemy force may ignore your hard work and simply take another route. In the twentieth century stationary defensive positions became fixed targets for long-range artillery and aerial bombardment. Defensive warfare became as mobile as offensive warfare and there was no longer use for permanent fortification. Modern technology has so favored offensive weaponry in recent years that defensive warfare no longer exists as such. Nuclear warheads, intercontinental missiles, and suicide terrorism leave everyone open to attack at all times, and there is no hiding behind picket fences, stonewalls, Maginot Lines, or missile shields. There are no physical barriers to modern weaponry. The only defense is the deterrent power of threatened retaliation or, in other words, more offense. Modern military technology is becoming so widespread in the twenty-first century that the idea of securing and defending any portion of the Earth's surface against any other portion is an absurdity. There is no such thing as defense. We may speak of "national defense," or "the Department of Defense," but there is no defense. There is no outside. There is only inside, and we are all in it.

Bad people do not come the way they used to come, in the aggregate. Walls do not separate us from them. They are in our midst and security is an internal problem. They are us, and with us behind the wall of gravity, and we can protect ourselves from them only on an individual, case-by-case basis, as we have learned to do.

Telegraphy

In his play *Agamemnon,* Aeschylus (525-456 B.C.) claims that the night Troy fell to the Greeks, their families hundreds of miles away knew all about it. As the city was captured by the likes of Agamemnon, Menelaus, Achilles, and Odyseus, a beacon was lit on Mt. Ida, about 100 km distant from Troy. A second fire relayed the signal west 154 km to the island of Limnos in the middle of North Aegean. A third fire kindled at Limnos was then seen 70 km distant at Athos, where a much larger fire was built that could be seen 177 km away on Euboea, an island off the mainland of Greece. From here the signal was relayed by the same means to Messapius, Cithaeron, Mt. Aegiplanetus, the Arachnaean height, and finally to the capital, Mycenae,

600 km. from Troy. This is how Clytaemnestra received tidings from her husband, King Agamemnon, whom she had not seen for ten years. (She had not been entirely waiting for him, as it turns out, and was not overjoyed to get the call.)

This may have happened. There is no corroborating evidence that it did, but under good conditions large bonfires lit on the peaks described can be seen across the distances described. The chances of everything working as planned, however, would have been minimal. Fog, rain, or an inattentive sentry at any point along the way would have ruined the whole thing. Even if the signal did get through, the information it conveyed would have been extremely limited. The message would have to have been pre-arranged. It could in no way describe *what* happened, only *when*. It could not describe any variation on the pre-arranged information and there was no way to reply to it or confirm its reception. Nonetheless, Agamemnon's message to his wife 3000 years ago, however crude it may have been and whether or not it happened, was one of the only ways human experience was transmitted rapidly over great distances in ancient times. Signaling was the beginning of telegraphy and telephony and television, the beginning of seeing through the eyes of others many miles away. Clytaemnestra was at home, in Europe, but knew what was happening in Asia.

In the ancient world there were two ways to send messages: by prearranged signal as described by Aeschylus, or by courier. A beacon fire, a flag or a mirror reflecting sun-light delivered information of low value, but it was fast. A courier, on the other hand, could deliver information of high quality, but it was slow. Stories of heroic messengers abound in ancient literature. We all know of Phidippides who ran 26 miles from Marathon to Athens in 490 B.C. to warn of an approaching Persian army. Until two centuries ago, this was still the best way to move information. Every ancient empire, every medieval kingdom, and every modern nation-state has employed couriers, both horse and human, from the Assyrians and Persians to the Incas and Romans, right up to the American Pony Express of 1861.

The written word lends itself well to courier service. Its greatest asset is accuracy. Once a message is solidified in material form, it may move undistorted virtually indefinitely through space and through time. An order, a description, an answer, or a question written here and now can be carried to any distant location and read, exactly as written, at any time. Much of what we know of ancient times comes of our eavesdropping on written messages that have traveled unmolested through time. But time is also a great limitation. Old information is no information at all, and couriers are slow, no matter how fast they are. Writing confined to a material medium (a piece of paper, a stone tablet, or an arrangement of beads on strings) is readable at distant locations only when transported physically. Great distances mean great time lapses, and a rapid deterioration in the value of information delivered. By the time they get to where they are going it is usually too late. If you wait a week for a response, you do not feel that you are conversing. True telegraphy, or messaging at a distance more or less instantaneously, requires freeing language from matter, and thus from the time lag associated with inertial mass. It means moving experience through physical space without moving anything physical. It is seeing what someone else sees, while they are seeing it. This was not possible until modern times.

An exception is an instance of true telegraphy reported by the historian Polybius (ca 200-118 B.C.). A "torch telegraph" was developed and used, he claims, by Aenias some 200 years before his own time: Two torches were raised and lowered at even intervals to represent letters of the Greek alphabet. The first torch was raised one to five times, indicating one of five predetermined divisions of the alphabet. The second torch was then raised to distinguish one of five letters within that division. The two torches, each with five possible signals, created a total of 5 x 5 = 25 possibilities, enough to represent the 24 letters of the Greek alphabet. Philip of Macedon (238-179 B.C.) reportedly used the same or a similar system. Later, the Roman general Sextus Julius Africanus (200-250 A.D.) used a three-torch system in which each torch was raised one to three times, for a total of 3 x 3 x 3 = 27 possibilities, more than enough for the Roman alphabet. Modern trials have shown these systems to be effective up to 8 words per minute. But they depend on clear weather, nighttime conditions, and extreme care in transmission and reception. A torch has to be raised high enough for a viewer to distinguish it from an unraised torch, and it has to be raised far enough away from other torches for the viewer to tell which one it is. Errors occur frequently and must have been rampant in ancient times. With care and patience any message can be spelled out over a long enough time, but the system is severely limited by the distance over which it can be seen clearly. A massive beacon fire can be seen for 100 km, but a torch is much smaller. The viewer must see it *and* see the space between it and the next torch. The message can, of course, be received and retransmitted to the next hilltop, but time is lost with each relay and the possibilities for error are multiplied. We do not know how often the torch telegraph was used or how effective it might have been, and there is no record of it being used on a permanent basis at fixed locations. We are reasonably sure that it was actually used, however, and it remains the first system we know of that encoded intelligence and that was sent both ways over distance.

Greater accuracy and information content was achieved in Europe and elsewhere in modern times with mechanical signals such as shutters, flags, and semaphores. But line-of-sight telegraphs, as they were known, were useful only during clear weather and daylight hours and only over distances even shorter than required for the torch telegraph. Signalers communicated over short distances between stations on towers, ships, or mountaintops. Longer distances proved impractical, as the relaying of signals from station to station usually resulted in a marked deterioration of accuracy. Couriers remained the only reliable form of intercity communication until the nineteenth century.

But in the half-century before the coming of the electric telegraph, a line-of-sight semaphore telegraph flourished briefly but gloriously. In the late 1700s and early 1800s an entire new generation of long distance mechanical semaphores was made possible through the use of telescopes. The telescope was invented in 1608, but was not manufactured in large numbers or in usable form until after 1757 with the introduction of the achromatic object lens. Telescopes reduced the number of interval stations over which a signal had to be relayed by vastly increasing the distance over which they could be read. Fewer stations meant increased speed and accuracy, to the extent that permanent intercity signaling systems were built in many European countries. The first and greatest semaphore telegraph system was designed, built, and operated in France by Claude Chappé.

On July 26, 1793 the French revolutionary government appropriated 58,400 francs for the construction of a mechanical telegraph line of fifteen stations from Paris to Lille, about 190 km north at the frontier of the Austrian Netherlands. Chappé was given the title *Ingenieur Telegraphe* and assigned to build the new system. France was at war with nearly all of Europe at the time, and the military importance of communication between the capital and a frontier post went unquestioned. Chappé had the system up and running within a year.

Each station along the line was placed on a hilltop, a tower, or a belfry, about 10 to 12 km from stations before and after. Trees were removed to improve visibility. An iron mast 5 meters high was erected at each location with a 4.5 meter relatable wooden cross beam or *regulator,* mounted at its top. The regulator had four settings: vertical, horizontal, left diagonal and right diagonal. To increase the number of settings, two *indicators,* or two-meter long rotating beams with a lead counterweights were mounted at each end of the regulator. The regulator and two indicators simulated the outstretched arms and flags of a semaphore signalman, and the system became known as a "semaphore" telegraph. Each indicator could be spun to eight possible settings at 45 degree angles from one another, which brought the total of possible settings for regulator and indicator to 4 x 8 x 8 = 256. This was reduced to 4 x 7 x 7 = 196 because one of the indicator settings was in line with the regulator and not easily distinguishable. After the system had been in operation it was decided to set the regulator only at vertical and horizontal positions, further reducing the number of settings to 2 x 7 x 7 = 98. These were later reduced to 94. Ten of these settings were reserved for the numerals 0-9, 26 for the letters of alphabet, and the rest for frequently used words and syllables.

The regulator and indicators were operated from below the mast with a control bar about a meter long that always stayed parallel to the regulator. When an operator grabbed handles at each end of the bar and turned it horizontally,

diagonally, or straight up and down, ropes extending up-
ward moved the regulator to the same position. This process always began a signal setting. Next, the operator used
cranks at both ends of the control bar to turn the indicators
to their proper settings. After some use it was realized that
information was transmitted more swiftly and accurately
if the settings were kept as simple as possible. Later versions of the codebook always kept the numbers of settings
to 94, but increased the number of settings per signal, so
that a single word or syllable might be represented by two
or three settings in sequence. The new codebook included
94 pages, each page with 94 entries denoting a commonly
used word or syllable. The first "division" of signals was
the original settings for numbers and letters and was used
when words had to be spelled out. The next two divisions
were for more common words and syllables. Decoders presumably turned to a page between 1 and 94 when the first
setting was sighted and then to an entry between 1 and 94
on that page in order to read the signal. With three divisions, the complete code then consisted of 94 x 94 x 94 =
8930 possibilities. Further divisions of the codebook were
added later.

Security was of primary concern, as much of the information passed down the line of telegraph stations involved military intelligence. The code was changed constantly, probably at the transmission of each message. The
code used for a particular message was itself encoded at
the beginning of each transmission. Very little information

survives as to what codes were actually used, as security precautions required that very few books be distributed. It seems that tilting of the regulator beam 45 degrees to the left indicated which division of the codebook was used; tilting to the right indicated control signals used by the operators themselves to keep the system in service. (Diagonal settings of the regulator were originally part of each signal setting as mentioned above; most likely it was found that using them only for special meanings outside of the message itself was a simplification that allowed for smoother operational control.)

Signals were passed down the line one at a time, each station receiving and sending them on to the next station as quickly as possible. One operator at each station was kept busy with two telescopes, watching the station behind him for new signals and the station ahead to confirm that the signal had been relayed successfully. He verbally relayed the position of the station behind him to an operator at the control bar of his own station, who then cranked the semaphore into the same position. As it was relayed down the line he recorded the setting in a logbook in the form of a pictogram, but he had no way to decode it and did not know what it meant. As soon as he saw that the station ahead of him was displaying the correct setting, a confirmation signal was sent back to the station behind. Mistakes were indicated by an error signal and corrected before the transmission continued. Operators were trained to be extremely attentive and precise in their actions, but were never allowed to know what they were saying to each other. The only codes they were permitted to read were the control signals that applied only to operational procedures and not to the messages conveyed. The message was decoded only at division and terminal stations and known only to a very few people.

An enormous public relations boost for Chappé's system came as the first official transmission from Lille reported a French victory over the Austrians and Prussians at Le Quesnoy. The message arrived in Paris a few hours after the victory. This was a miracle. Parisians were used to thinking of Lille and any information coming from Lille as two or three days distant; to know what had happened *that day* was astounding. Telegraphed reports of other victories soon followed and Chappé became a hero. The new semaphore system seemed to confirm the principles of scientific and social progress that underlay the French Revolution. The government, now able to know what its armies were doing, virtually as they were doing it, soon authorized new telegraph lines throughout the country. The telegraph developed into a nationwide network linking Paris with 29 cities through France and neighboring countries.

Five principle lines radiated from the capital. Four of these terminated at the four faces of the headquarters tower of the Telegraph Administration in Paris. (One of them split into two separate lines at the next station.) Here outgoing messages were sent to all parts of the country and incoming messages were decoded and delivered to government ministries. The five principle lines terminated at hub stations in Lille, Metz, Lyon, Bordeaux, and Avranches, where messages were decoded and retransmitted on connecting branch lines to more distant cities. Fifty stations connected Paris and Strasbourg. Napoleon was especially thrilled by his ability to keep in touch with his armies throughout Europe by means of the Chappé system. While in Spain he was informed that the Austrians had invaded Bavaria, and was able to meet the enemy there four days later. In 1809 he wrote to the Minister of the Interior: *MrCretet, I wish that you complete without delay the establishment of the telegraphic line from here to Milan, and that in fifteen days one can communicate with that capital.*[28] He later commissioned Chappé's brother Abraham to develop a mobile version of the telegraph for use during the invasion of Russia. At Boulogne, a station was built capable of signaling across the channel in preparation for Napoleon's planned invasion of Britain. By 1852 lines were extended to Spain, Italy, Germany, and Belgium. Over 4000 km of telegraph lines were built in all, including 556 stations.

At least two operators were needed at each line station and five or six at each terminal station, including operators, reserve operators, inspectors, and engineers. In all, the system employed over 1000 people. It took each station twenty to thirty seconds to transmit a signal. Reports of end-to-end transmission times vary, and speeds were often enthusiastically exaggerated. Under optimum weather conditions, a report by Rollo Appleyard states that *"At Paris, when this system had been perfected, they could receive communications from Lille in two minutes, from Calais in four minutes five seconds, from Strasbourg in five minutes fifty-two seconds, from Toulon in twelve minutes fifty seconds, from Bayonne in fourteen minutes, and from Brest in six minutes and fifty seconds."*[29]

This probably applied to a single signal. Complete messages, sent one signal at a time, took considerably longer. In bad winter weather messages often took several days to get through. But, as supporters of the telegraph often pointed out, couriers were even more likely to be delayed in bad weather.

Chappé's success soon led to the construction of mechanical line-of-sight telegraph systems in Sweden, England, Norway, Denmark, Finland, Germany, Belgium, Netherlands, Ireland, Spain, Australia, Russia, and the United States. Inspired by the French system, Abraham Edelcrantz built a similar system using 50 stations over 200 km of southern Sweden. Instead of the semaphore, Edelcrantz employed a system of ten stationary shutters that, left open or closed, relayed information by patterns chosen to relay letters and numbers. Unlike the French, who allowed original messaging only between terminal stations, the Swedish system allowed messages to be transmitted between line stations so that smaller cities were able to communicate with the capital and with each other. The British, under Lord George Murray, developed a nine-shutter system for the admiralty in 1796. The line stretched along ten stations from London to Portsmouth. In good weather a message from the Admiralty office took 15 minutes to reach the fleet in Portsmouth. Other lines were later built from London to Deal, Sheerness, and Yarmouth. The system worked effectively to coordinate naval preparations during the Napoleonic Wars, but was dismantled after the wars and replaced by a semaphore system that was used until 1851. In the U.S. shorter lines were built from Boston to Martha's Vineyard and between Boston and Nantucket. In San Francisco a three-station system was constructed linking "Telegraph Hill" to Presidio and Point Lobos.

Mechanical telegraph systems were considered the wave of the future. Edelcrantz, founder of the Swedish system, wrote in *A Treatise on Telegraphs* (1796) that *"By increasing the number of shutters of the telegraph and the number of signal combinations, so that the signals can express ideas instead of syllables and letters, it would be possible to express one or several meanings with one signal, not predetermined as they are at sea, but containing all the turns and variations in ideas and expressions occasioned by each new circumstance. One might even be able to create a kind of universal language, whereby people of all nations could communicate without knowing each other's language."*[30]

But with the coming of the electric telegraph, the mechanical systems fell out of use. The last successful military use of mechanical semaphore came in 1855 on a semaphore line from Moscow to Sevastopol during the Crimean War. The line was in place when the war began, and greatly assisted Russian military authorities in coordinating troop movements. When the war started, British and French forces relied on steamships, couriers, and regular mail service across Bulgaria and the Black Sea to keep in contact with their capitals, a process that took two or three weeks. As the war continued, they laid an underwater electric telegraph cable across the Black Sea from the Crimean Peninsula to Bulgaria and connected it with the trans-European telegraph network to Paris, and thence across the Channel by another underwater cable to London. Late in the war the Russian mechanical semaphore was replaced with an electric telegraph cable.

What might it have been like to be an operator of the Chappé system, say on the line between Paris and Brest? At one of the several dozen stations along the way you might notice through your telescope that the station ten kilometers to your west, toward Brest, has tilted its regulator to one side, preparing to send a signal. You immediately alert your co-worker, who jumps to the control bar of your station and tilts the regulator in the same manner. You write down the time. Turning to the east, you glance through the other telescope, watching the station in that direction (toward Paris) as it confirms that it has received your alert signal. You switch back to the west, waiting for the signal to develop. There is a pause. As you watch, the regulator turns suddenly to a distinctly horizontal position. The left indicator turns out and down 45 degrees and the right indicator points straight up. You write these positions down in your logbook as you shout them to the man at the control bar. He muscles the heavy iron regulator to the horizontal position, and then wheels the right and left indicators into positions you have seen and relayed to him. He shouts back that his signal is complete. You watch the station to the east, toward Paris, as it repeats the signal, then you turn west again, watching for the next signal from Brest. The regulator in the distance swings upright and back to the horizontal. You shout this down to the operator as you watch for the new indicator positions. The left indicator turns up and in 45 degrees, and the right indicator points down and in 45 degrees. You write this down as you convey it to your controller, turning to confirm an accurate reception at the next station. Assured that the information is repeated correctly, you look for the next signals. They continue steadily for twenty minutes and then the "end" signal comes. You relax, but know that a reply from Paris is likely within the hour.

What was said? You have no idea. Only the "engineers," at each terminal and a few others have access to the codebook. You know that you accurately relayed signals through your station from west to east, but how, you wonder, do they "say" anything? How is it that something the transmitting officer says or sees or thinks is transferred to the receiving officer hundreds of miles away through an iron crossbar and two wooden sticks? How are mental processes conveyed through a purely material medium? What you see through your telescope is identical to what the receiving officer sees through his, but he knows what the transmitting officer sees or thinks while you do not. He has access to an entirely separate realm of conscious experience.

Your experience is limited to perception, or what you see for yourself directly. His experience is perceptual (he sees what you see), but he also sees "through the eyes" of the person transmitting the signal. His experience is *reducible* to perception in that all he actually sees are the same semaphore settings you see, but the particular order in which those settings are relayed creates a realm of consciousness beyond direct perception. Unlike you, the receiving officer at the terminal in Paris knows that a British ship has been sighted off the coast at Brest. He did not see the ship, but he knows it is there. Someone else saw it and encoded the message that you relayed to him. But how did this experience arise from iron and wood? To create it there must be a *potential*, or a limited number of possibilities within which signals can occur. There has to be a context within which each signal has meaning. In the case of the Chappé system, the potential is the settings of the regulator (2), left indicator (7), and right indicator (7), for a total of 98 settings (later reduced to 94). Every *actual* signal must be only one of these. An iron bar with wooden sticks on each end can have meaning only if arranged within a set of possibilities. There has to be a choice: some possibilities have to be actualized and others left empty. The settings *not* assumed by a signal are, therefore, as important as those that *are* assumed. There is no information conveyed unless some possibilities are left empty. The empty space surrounding the regulator and indicators is the potential; the particular message conveyed is the space that is filled.

"Seeing through the eyes" of others goes back as far as language, and perhaps farther. Words and letters convey consciousness only because they *could be* other words and letters. Like telegraphic symbols, they are particular choices within a potential. What makes telegraphy new and different is that real time human experience is suddenly raised from the level of the dining room, the street, and the village square to the level of an entire continent. People in Paris suddenly know about an English ship off the Atlantic coast, *while it is off the Atlantic coast.* This is a huge leap. In the case you experienced at the line station, it may be that the government knows of an impending British diplomatic mission to Portugal, but does not know when it is scheduled or if the envoy is on this ship. An intelligence officer reads the message and immediately sends a dispatch back down the line asking for a more detailed description of the ship. But there is barely enough time before it disappears over the horizon. At your post, you are looking alternatively east and west through your telescopes, waiting for the next signal. You have been trained to act quickly when a signal arrives. As expected, the signal arrives from Paris, from the station to the east. Its regulator swings to a diagonal, and you shout the alarm to your control operator. As the signal develops, you know that there is something more going on here than meets the eye.

As line-of-sight semaphore systems were springing up across the continent scientists were beginning to experiment with electrical telegraphy. Even Claude Chappé toyed with electric telegraphy in the 1790s before giving it up in favor of the mechanical system. Early devices were electrostatic or electrochemical. As the name implies, electrostatic telegraphy did not employ a completed circuit to deliver a signal. Instead, charges were generated in Leyden jars and transferred by copper or iron wires to a row of lightweight pith balls suspended on strings. Each ball stood for a letter of the alphabet and was deflected by the electric charge in a particular wire. A transmitting operator chose the letter he wanted and deflected the pith ball by connecting its particular wire to the Leyden jar. The system worked over short distances, but was slow and required the extension of twenty-six wires between stations. Electro-chemical devices were no faster or less cumbersome. They were based on the electrolysis of water, or the visible separation of oxygen from hydrogen in the presence of an electric current. A transmitting operator caused gas bubbles to appear in twenty-six separate enclosed glass containers at the receiving end by connecting wires corresponding to letters of the alphabet. This system was also limited in the distance over which it could generate a signal. Another interesting system was neither electrostatic nor electrochemical: the operator received signals by feeling electric shocks through wires attached to his fingers! These early systems provoked interest in the scientific community, but were considered by the general public to be little more than curiosities. Until the 1840s, none of them competed seriously with the mechanical semaphore.

The breakthrough in electric telegraphy was electromagnetism. Hans Christian Oersted discovered in the early 1800s that an electric current passing through a wire would deflect a nearby magnetic needle. This was a fundamental scientific discovery. It meant that moving electricity creates a magnetic field, which meant that magnetism could be controlled at a distance and thus the movement of metal objects could be controlled at a distance. By manipulating the flow of electricity through a wire at one location, one could control a magnetic field at another location, and thereby move an iron armature at will. More importantly from the standpoint of communication, one could decide *not* to move it. One could cut off the electricity and leave the armature where it was, or turn it on and then off, or leave it on for a while and then turn it off. The possibility of moving or not moving the armature created a potential for the flow of conscious experience between one observer and another. (A permanent magnet is of no information value whatever, as it cannot *not* move the bar.)

A number of obstacles remained for the practical development of an electromagnetic telegraph. In 1824 an English mathematician by the name of Peter Barlowe experimented with electromagnets at various distances and found that the current decreased with the square root of the length of the wire. The signal became too weak at more than 200 feet, and he was forced to conclude that electromagnetic telegraphy was impractical and theoretically impossible. His problem was a lack of theoretical knowledge as to how electricity could be shaped and sized for particular purposes. Barlowe used a simple battery with a single pair of metal plates immersed in an acid solution, or what we would call a single-cell battery. In 1829 and 1830 an American scientist in Albany, New York by the name of Joseph Henry, lined up a series of battery cells, each with its own pair of metal plates, and began experimenting with electromagnets. When all the negative electrodes were connected to one wire and all the positive electrodes to another, he had what he called a "quantity" battery, or one that produced a large amount of current and a powerful magnet. We would call this a parallel battery circuit, which is essentially one large battery. But when he connected the negative electrode of one battery to the positive electrode of the next he created what he called an "intensity" battery that, though no more powerful than a single cell, could carry a current a much greater distance. We call this a series battery circuit, or one that produces the same *amperage* (current) as a single cell, but greater *voltage* (pressure). Enough battery cells linked in series could produce magnetism at almost any distance. But the most significant practical result of Henry's experiments was the use of intensity and quantity batteries in combination. An intensity battery could be used at a great distance to move a very small iron bar a very small distance to act as a switch. The switch could close the circuit of a nearby quantity battery that could move something much larger, say, the clapper of an alarm bell. Alarms became essential in notifying telegraph operators that a signal was forthcoming. Or, the switch could be used to turn on another intensity battery that would step up the strength of the original signal. This became known as an electric *relay*. Relay stations, each with its own intensity battery, could be installed at fixed distances along a telegraph line. Henry's experiments with electromagnetism produced the theoretical possibility of magnetism of any strength delivered to any distance.

Samuel Morse stepped onto the stage in 1837. He was a professional studio artist for many years and recently employed as professor of the Literature of Arts and Design at the University of the City of New York. Morse was not a scientist and had little or no knowledge or understanding of advances in electrical telegraphy then current in Europe, and was entirely unaware of Henry's experiments in Albany. Intrigued by the possibility of creating telegraphy "if the presence of electricity can be made visible in any desired part of the circuit,"[31] Morse had been experimenting for several years with a recording telegraph. He considered the insights he developed entirely original to himself. In his system, a piece of paper moved steadily beneath an electromagnetic armature that deflected a pencil sideways. When at rest the magnet left a straight line; when activated it left a V-shaped jag in the line. A message was sent from one end of a wire to the other coded in the number of consecutive Vs drawn by the pencil. (The line was allowed to go straight between each numbered signal.) Instead of corresponding to letters of the alphabet, each number corresponded to a whole word. Morse developed a codebook for his system by matching numbers to words in a dictionary.

Morse's system was ingenious but useless. Like others of its time it worked only over short distances. Without distance there is no telegraphy. The only real advantage it had over earlier systems was that it required only two wires. The decisive moment came for Morse when he showed his machine to a colleague at NYU, a Dr. Leonard Gale, who was familiar with Henry's work. Morse and Gale formed a partnership and soon incorporated a twenty-cell intensity battery into the system. They could now send signals over great distances. In October of 1837 Morse filed a preliminary application for a patent. Another colleague, Alfred Vail, was brought into the partnership and improved the Morse system by reworking the armature to move up and down rather than back and forth. The new vertical armature produced dots and dashes on the moving paper instead of Vs. This was the celebrated "invention" of the telegraph. It may have been Vail who dropped the numbered word system and instead used dots and dashes to represent individual letters. In 1844 a 40-mile long telegraph wire was strung overhead between Baltimore and Washington D.C. and the Morse system became a working reality.

By that time several non-armature telegraph systems had been invented in Europe. A six-wire system used electromagnets to turn magnetic needles to the left or right by reversing polarity of the electricity. Each of three needles could be turned simultaneously to one of three settings:

right, left, or neither, producing 3 x 3 x 3 = 27 possible settings representing letters of the alphabet. But the system, though successful, remained a demonstration and was never placed in practical use. Cooke and Wheatstone developed a long-distance system in England after Joseph Henry brought his discoveries to England in the spring of 1837. One of their systems used five needles, each with its own wire and each deflected in only one direction. Five needles deflected simultaneously, each with two needle positions (deflected or undeflected), produced 2 x 2 x 2 x 2 x 2 = 32 signal combinations. Other needle systems, including a 25 letter system later developed by Cooke and Wheatstone and their competitors, several were put into practical use along English railway lines in the 1840's. A dramatic demonstration of the usefulness of the electric telegraph occurred in 1845 with the arrest of John Tawell, who was later convicted of murder. Seen boarding a train at Slough, his description was telegraphed ahead to Paddington where he was apprehended. The actual message, which must have been through the 25-letter system, described Tawell as "dressed like a Kwaker."

But needle systems could not compete with the electromagnetic armature. The Morse system prevailed in America and later in Europe because of its simplicity. The limiting factor in all systems was never the ingenuity of the sending and receiving machines, but the maintenance and security of the circuit between them: The fewer wires, the more practical the system. It was soon discovered that through use of an earth ground, the Morse system could be reduced to one wire. Batteries at each end of the line were connected directly to damp soil, obviating the return conductor. Another advantage of the Morse system was an unintended consequence of Vail's vertical armature. The old back and forth armature developed by Morse produced no sound, but the vertical armature produced an audible "click" as it snapped into place. Morse intended that the message be recorded on the moving paper and translated only after the transmission was complete. He considered the sound of the armature an annoyance. But operators found they could read the message audibly *as it was transmitted*, and simultaneously translate it into written or oral language. Communicating entirely with each other, they did not need to write at all. Electrical communication was actually faster than writing. This enhanced the real-time effect of electric telegraphy and created the phenomenon of two-way long distance conversation.

To what extent, then, was Samuel Morse the "inventor" of the electric telegraph? Only, I think, to the extent that we require an inventor. Plainly, others contributed as much as he, and more plainly still, the telegraph was on its way no matter who ended up with the credit. We require a story for our inventions, a myth of origin, an odyssey of the trials and labors of a single person. We need this in order to believe that consciousness is contained by individuals and that ideas come from the minds of single people. We understand the possibility of collaboration, but we have difficulty

perceiving the wholeness above and beyond individuals. It
is this that Morse personifies. He was not the inventor, but
he was the one who drove the idea through to completion.
He secured the patent and made it a commercial success.
As Joseph Henry stated: "I am not aware that Mr. Morse
ever made a single original discovery, in electricity, magnetism, or electro-magnetism, applicable to the invention
of the telegraph. I have always considered his merit to consist in combining and applying the discoveries of others
in the invention of a particular instrument and process for
telegraphic purposes."[32]

The need to identify inventions with specific individuals led to great anguish for both Chappé and Morse, and
for many who worked with them. Chappé, suffering from
depression, jumped down a well in 1805 after many chal-
lenges to his primacy as inventor of the semaphore telegraph. Morse was challenged more systematically, in print
and in court. He claimed to have conceived the electric
telegraph in discussions with fellow passengers on board
the ship *Sully* while returning from France in 1832. One
of the passengers, Charles C. Jackson, later claimed that
the telegraph was "our invention" (though, unlike Morse,
he never tried to develop it.) Morse later suspected that it
was through Jackson that "his" idea leaked out to Cooke,
Wheatstone, and others working on telegraphy in Europe.
(Cooke and Wheatstone later fell out with each other in
England.) How else could they have known what *he* was
thinking? Many of his business partners and associates,

including Francis O.J. Smith, Henry O'Reilly, Ezra Cornell, and Cyrus W. Field, bitterly challenged his patent rights and exclusive claims. Alfred Vail complained for years that Morse did not recognize his contributions, and Vail's widow contested Morse's "invention" long after the fact. Joseph Henry, though a supporter of Morse at first and never interested in the commercial application of the telegraph, also fell out with him later on, feeling that his seminal work on electromagnetism was not recognized. Morse thought of himself as the lone inventor, who had conceived the telegraph wholesale and brought it into the world once and for all. Improvements by others in later years he considered little more than tinkerings. In 1858 he was instrumental in laying the first transatlantic cable from Ireland to Newfoundland, but considered it an idea that no one had thought of before him.

Morse was more than a compiler of other people's work. Despite his lack of scientific background, he did manage to develop an expertise in electromagnetic signaling and did develop a primitive system of his own. A born promoter, he gave numerous open demonstrations of what he achieved, and managed to link his name with the invention in the public mind. Everyone admired what he had done, but for many years no one had a use for it. He applied and re-applied for congressional appropriations to build an interurban line, but funds were not forthcoming. Morse and his partners soon turned to other pursuits. Ultimately, it was not his ability to focus on the telegraph that led to success, but his inability to focus at all. From the time he first demonstrated the system in 1837 at NYU to the completion of the Baltimore to Washington line in 1844, Morse, still a professor at NYU, spent eleven months in Europe soliciting for patents, retained a commission to continue his painting career, was re-elected president of the National Academy of Design, opened one of the first photography studios in the United States, and ran for mayor of New York. He worked on telegraphy all the while, but had he concentrated enough to succeed at any of his other endeavors he would likely have forgotten all about it.

In 1843 Congress finally came through with an appropriation for a telegraph line. Morse's partners, Gale, Vail, and Smith, were no longer actively promoting the invention, and had scattered to New Orleans, New Jersey, and Maine. It is at this point that Morse earned his place as father of the American telegraph. It was he who brought the partnership back together to secure right-of-ways, purchase cable and poles, prepare batteries, relays, and transmitters, and coordinate operations with Congress. The line was ready by June of 1844, just in time for the Democratic Party convention in Baltimore. Vail was stationed in Baltimore and telegraphed the breaking news of James Polk's nomination to Morse, who had set up his station in the Supreme Court chambers (then situated in the capitol building). Congressmen and senators waited anxiously outside the door as Morse relayed the results, ballot by ballot. The versatility of the new machine was demonstrated the next

day when the convention in Baltimore telegraphed its nomination of senator Silas Wright for vice president. Wright
immediately wired back from Washington declining the
offer. A second message asked him to reconsider and a
second reply stated that his mind had been made. Further
messages flew back and forth between the two cities, each
"with *lightning speed*," as reported by the *National Register.*
The magic of instantaneous communication over long distances became a sudden reality to people in a position to
make things happen. Morse was a hero.

Reactions in the popular press were a mixture of awe
and bewilderment. "We stand wonderstruck and confused,"
said one paper. Morse had "annihilated space and time,"
created a "new species of consciousness." The human mind
could be suspended in wires and batteries; people in far-
off cities could converse as if in the same room. "This extraordinary discovery," one newspaper reported, "leaves…
no elsewhere – it is all *here*." The New York Sun called
the telegraph "the greatest revolution of modern times and
indeed of all time, for the amelioration of Society." Civic
ties would be strengthened coast to coast, fugitives would
not escape the law, absent relatives would barely seem away
from home, and nation would no longer rise against nation. There would be a single nexus of knowledge and understanding. Most Americans had not known for weeks
about the Declaration of Independence, and the Battle of
New Orleans had been fought three weeks after the War of
1812 was over, but now everyone would know everything

as it happened. "What a future!" proclaimed the *New York Herald.*[33]

The Morse system required up to four signals to send a single letter. Each signal of Chappé's mechanical semaphore carried much more information than the dot or dash of the Morse system. The older system had 94 possible settings of the regulator and indicators: A letter, numeral, syllable or even a whole word could be sent with a single setting. (This seems to be what the first electric systems were trying to emulate.) The Morse system had only two settings: dot or dash. But where the mechanical telegraph signal took 20 to 30 seconds to develop, the Morse system took a split second. Far more words per minute could be sent between one station and another. More importantly, the distance between stations could be increased from a few visual kilometers to virtually any terrestrial magnitude. The number of stations was reduced from forty or fifty to two, and the possibilities for error reduced accordingly. As electric telegraphy developed in the middle and late 1800s, means were developed for sending two and then four messages simultaneously over a single wire. Because messages could be sent and received faster than people could read and write them, it was soon discovered that they could be written and encoded "off line" by punching holes onto a strip of paper. The paper strips were sent in rapid succession through a machine that converted the holes in the paper to dots and dashes, automatically "keying" the message on line. A message that took a human operator minutes to tap out was sent in seconds. A single wire could thereby send dozens of messages every minute. The Morse system handled so much more information than the old Chappé system that its use shifted from primarily government and military to commercial and personal correspondence.

Telegraph wires between cities were laid both underground and overhead. They did not generally follow old line-of-sight telegraph routes, however, as these usually crossed hilltops and other inaccessible areas. Wires had to be within reach of work crews that monitored and repaired them on a continuing basis. A fortunate coincidence was the simultaneous construction, in both Europe and America, of the railway network. Telegraph lines were constructed conveniently along railway right-of-ways. Trees, buildings, and other obstacles were thereby avoided, rivers and mountains crossed, and ready access made available for maintenance. As information on the progress of trains entering and leaving stations became vital to maintaining passenger and freight schedules, railroads needed the telegraph as much as the telegraph needed the railways. This interdependence led invariably to symbiotic right-of-way contracts between railway and telegraph companies. Telegraph wires soon spread across Europe, the Middle East, India, South Africa, Siberia, China, Australia, and America. Underwater cables spanned the English Channel in 1851, the Atlantic Ocean in 1866, and the Pacific Ocean from Australia to Victoria BC, in 1902. A final segment from Fiji to Fanning Island was completed later that year and a telegram was sent around the world from England through South Africa, Australia, and Canada and back to England. In 1903 another Pacific cable was laid from the U.S. to China and the Philippines.

The electrical telegraph is actually a step backward to a material medium for long distance communication, to the extent, at least, that an electron is material. It is slightly slower than the semaphore to the extent that an electron moves more slowly through a copper wire than light between semaphore stations. But an electron moves so fast that the time delay at any terrestrial distance is imperceptible. Transmission and reception are virtually simultaneous; the experience of the receiver becomes the real-time experience of the sender. This quality of electricity, more than the quantity of information it carries, is its revolutionary contribution to the expansion of human consciousness.

More important even than the fact that one operator could talk to another over distance in real-time was the fact that he could do so while bypassing what might be called the "perceptual sphere" of individual observers. By "perceptual sphere" I mean the few meters distance within which an individual normally sees, hears, smells, tastes, and feels. A room in a building, a comfortable shouting distance, or at most a visual horizon is a perceptual sphere. It is the realm of individual experience familiar to each of us, and the limit of direct interpersonal communication. The mechanical semaphore was a vast extension, through the telescope, of the perceptual sphere and a linking over great distances of one sphere to another. It was a linear chain of personal distances stretching from one end of a continent to another. Its message was the telling and retelling of the same story from one person to the next, amounting to a highly sophisticated word-of-mouth transmission from sender to receiver. Electricity linked sender and receiver directly. Where Chappé's telegraph carried an impulse from one human to another the way a tactile impulse is carried incrementally through a body of cells, Morse's telegraph interconnected distant humans the way a nerve fiber connects one distant cell to another. And it uses the same medium.

Reproduction

In December 1606, three English ships set sail for Virgina under the leadership of Captain Christopher Newport. After prolonged adventures in the Caribbean, they landed at the mouth of Chesapeake Bay. Searching for a suitable site to found a colony, they worked their way up the James River and built a triangular fort on Jamestown Island, "a verie fit place for the erecting of a great cittie," according to Captain John Smith.[34] Site selection was not for the sake of economy, agriculture, or health, but for defence: defence more against European enemies – the Dutch, the French, and especially, the Spanish – than against the indigenous population. The island (sometimes a penninsula, depending on the vississitudes of the river's silt deposits) provided a deep water port on one side and defensive advantages, as it was a safe forty miles upstream from the open ocean. But it was swampy, mosquito ridden, and too small to provide room for hunting or sizable farming. Its major disadvantage was lack of good drinking water. Within weeks the colony was beset with fevers, infections, dysentary, and Indian attack. In January the fort burned down. Of the 105 men who landed in May, 1607, 38 were alive eight months later.

The colony would have died out in short order, had not hundreds of new colonists arrived in 1608 and 1609. But new settlers were not put to good use and the colony did not learn to produce its own needs. It did not even feed itself. Corn was grown in small patches, but settlers depended on re-supply from England and on trade with the Indians. Favorable trading terms for corn, venison, and turkey were assured by manipulating the Indians' fear of muskets and cannon. Iron tools, copper, and beads were given in exchange. In 1609 Wahunsunacock, chief of the Powhatan Confederation, put a stop to the trade when he realized that it was keeping the colonists alive as they encroached on Indian lands. Supplies from England were interrupted that same year by bad weather and shipwreaks off Bermuda. The winter of 1609-10 became known as the "starving time." Of 600 settlers, 60 survived. These were rescued with the arrival of Lt. Governor Sir Thomas Gates in May 1610. Gates "found the Pallisadoes torne down, the Ports open, the Gates from off the hinges... emptie houses rent up and burnt rather than the dwellers would step into the Woods a stones' cast from them, to fetch fire-wood; and it is true, the Indians as fast killing without, if our men stirred but beyond the bounds of their Block-house, as Famine and Pestilence did within." He brought the survivors aboard and decided to "quit the Countrye."[35]

The colony a total failure, Gates set sail for England. But before reaching the mouth of the James River, he met a fleet arriving from England with 150 new settlers under the command of a new Governor, Lord De La Warre. The survivors of the "starving time" were ordered back to Jamestown, and the English experiment in America was saved. Fortunately, they had not burned the fort as they left. Once the colony was re-established, the De La Warre administration went on the offensive against the Powhatans, killing, burning, and looting villages and crops. That spring 600 more settlers arrived, most of whom died over the summer. But more arrived later that year and the following year. By 1619 new arrivals had brought the population to about 1000. In the next three years, 3570 more arrived.

By March 22, 1622, 1240 were left. That day 347 of these were killed in an Indian Massacre led by Opecancanough, brother of Wahunsunacock. The survivors retreated to the fort, where faminie and pestilence killed 600 more. But the ships kept coming and the population recovered, doubled, and doubled again. Systematic destruction by the English of Powhatan villages, crops, and people ensued, no holds barred. In 1644 Opecancanough organized another massacre that killed 500, but by that time new settlers were arriving by the thousands. More importantly, they were raising families, and permanent English settlement of North America was assured.

Why did this happen? What were people looking for when they came here? Did they find it? Presumably not, as most were dead in a matter of months. Every goal stated for the the original settlement went unrealized. Nothing went as planned. Yet here we sit, 400 years later, amid streets and fields and skyscrapers that are the growth of Jamestown. It was not worth the price for those who paid it, yet here we are.

The stated goals of the colony were commercial, nationalist, and religious, but from the beginning the real moving force was commercial. The Virginia Company, sanctioned though it was with a royal charter from King James I, was a business and not a government operation. It was to explore the area of North America between the Spanish settlements in Florida to the south and the French settlements along the Saint Lawrence River to the north, and to extract wealth in whatever form it might come. Gold was foremost in everyone's mind. The idea of staying and making a home of the place, though no doubt present in the hearts of some, was not the moving force behind the colony. Most settlers wanted to make money and leave. They expected to sail up bays and inlets, make contact with native peoples, find out where the gold, silver, pearls, gems and jewels were, steal them, and go home. The Spanish in South America were their model. They expected to live off company supplies and barter with the Indians. For years they resisted the thought of working to produce their own food.

The lack of mineral wealth disappointed everyone. Captain John Smith continually refers in his writings to the good fortune of the Spanish in getting the good colonies first. The Powhatans not only lacked gold, they had no cities, no industry, no cattle, tea or spices. There was nothing to plunder. The best that could be gotten was corn and the occasional animal carcass. There were no profits for the company's stock holders to distribute. The first ship returning to England in 1607, The *Susan Constant*, brought a load of fools gold. There seemed no way the colony could fulfill its mission, yet the ships kept arriving with more people.

Settlers already there were not always glad to see newcomers. The ships brought provisions, of course, but people coming off the ships often brought disease as well. Newcomers had to be fed, defended, and taught the ropes. There was no inn or public house for them to stay in, even as late as 1623. A visitor in that year reported that "The new people that are yearly sent over which arrive here for the most part very Unseasonably in Winter, finde niether Guest house, Inne, nor any the like place to shroud themselves in at their arrivall, noe not soe much as a stroake given towards any such charitable worke soe that many of them by want hereof are not onely seen dyinge under hedges and in woods but beinge dead ly some of them many dayes Unregarded and Unburied."[36] The Virginians defended themselves from this charge by assuring that newcomers were distributed in private houses and, "As for dyinge under hedges there is no hedge in all Virginia."

Wealth, if it were to be had at all, would have to be produced by the colonists themselves. But what form of wealth? What could be produced here that could not be

produced at home? Several ideas were tried. Ironworkers
were imported, some of them from Poland. Mulberry trees
were planted for silk production and a glassworks was built.
Vintners were brought in from France, and grapevines
planted. A salt works was built. The Virginia company,
needing good publicity to keep the stream of immigrants
flowing, published a series of pamphlets extolling the op-
portunities now available in the new world. But the only
thing that the colony could sell in any quantity was tobacco.
Europe was developing a serious nicotine habit at the time
and was willing to pay top shilling for as much of the weed
as could be grown. But even this was an artificial economy,
in that tobacco was also grown in England. In 1620, to
assist the colonial economy (really, to create it) the English
government banned the commercial production of tobacco
in the mother country. The first to successfully produce the
type of tobacco popular in Europe was John Rolfe, who
later married Pokahontas, daughter of Wahunsunacock.
Rejecting the local variety grown by the Powhatans, Rolfe
imported Caribbean seed, and became quite wealthy as a
result. Soon, tobacco was grown everywhere, to the detri-
ment of every other commercial activity, and became the
actual currency of the colonial economy. The same visitor
who thought he saw dead bodies under the hedges noted
that after the massacre of 1622 there was little else to be
exported from Virginia: "For the Iron workes were utterly
wasted and the men dead, The Furnaces for Glass and Pots
at a stay and in a smale hope, As for the rest they were

had in a generall derision even amongst themselves, and the Pamphlets that had published here beinge sent thither by Hundreds wer laughed to scorne, and every base fellow boldly gave them the Lye in divers perticulers, See that Tobacco onely was the buisines and for ought that I could here every man madded upon that, and little thought or looked for any thinge else.[37]

Is this what so many struggled and died for? Even John Smith, ever the optimist and apologist for the colony, concluded that "This deare bought Land with so much bloud and cost, hath only made some few rich, and all the rest losers."[38] The Virginia Company, failing in its primary purpose of making money for its owners, had its royal charter revoked in 1624 and went out of business.

Nationalist goals for developing an overseas colony were hardly more successful. England, if it were to be a great nation, had to expand. With the defeat of the Spanish Armada in 1588, the English saw their chance to spread out across the Atlantic. A colony in Virginia would keep the Spanish from coming north from Florida and provide a port from which to prey on Spanish shipping in the event of war. And who could foresee what other national strategic advantages might be revealed in time? Everyone was still looking for any easy passage to the Orient. It was known that there were mountains to the west of the fall line in Virginia, but hope remained that the Chesapeake or some other bay would pass through the mountains and connect directly to the "South Sea." (Balboa's first view of the Pacific Ocean faced south from Panama.) Many still thought of the Americas as an obstacle to pass on the way to China and India (and would have preferred it were not there at all). English naval and commercial power was in the ascendant; if a passage through America existed, the English wanted to control it. If England was to be a great power, America had to be settled by the English. But as those of us who are descendants of Jamestown survivors well know, the idea of spreading English sovereignty in America eventually backfired.

Religious goals could be cited to justify the Jamestown colony whenever commercial and nationalist goals fell short. An objective of Virginia Company activities in America, according to a company directive, was "...propagating of the Christian religion to suche people as yet live in darkenesse and miserable ignorance of the true knowledge and worshippe." It would be the company's obligation to "bring the infidels and salvages, lyving in those partes, to humane civilitie."[39] Instructions prepared for Sir Thomas Gates before sailing for Virginia in 1610 stated that the conversion of the Indians to Christianity was "the most noble and pious end" of the colony. Any cost here on Earth was justified by the altruism and otherworldliness of the goal. No matter how many earthly lives were lost, eternal lives would be won. Evangelism justified not only English sacrifices in lives and treasure, it also justified any sacrifices the Indians might have to make along the way. "These people," according to Robert Gray, "are vanquished to their unspeakable profite and gaine."[40] In Thomas Hariot's description of Virginia he states that the Indians, "through descreet dealing and governement," would be brought to "the trueth, and consequently to honour, obey, feare, and love us."[41]

Donors in England provided many of the material necessities. A college for the education of Indian children away from their parents was established at Henrico, just up the river from Jamestown, and substantial sums given "for the bringing up of the salvage children in Christianity." Additional sums were available "for three discreet and godly young men in the Colony, to bring up three wilde young infidels in some good course of life." Another school for the Christian education of Indian children was later built downriver from Jamestown with donations from members of the East India Company. Both schools were destroyed in the massacre of 1622.

Though there are few recorded instances of actual conversion, an exceptional case is that of Pocahontas, daughter of Wahunsunacock, the Powhatan chief. She was well known and appreciated at the fort in Jamestown, having brought food to starving colonists on several occasions. She also, according to his own later accounts[42] saved Captain John Smith from execution by her father. Later captured and held prisoner, she learned English and converted to Christianity. Smith writes of "...how careful they were to instruct her in Christianity, and how capable and desirous shee was thereof, after she had beene some time thus tutored, shee never had desire to goe to her father, nor could well endure the society of her owne nation..." She openly renounced her countries idolatry, confessed the faith of Christ, and was baptized "Rebecca."[43] She married John Rolfe in 1614. There is no doubt as to her love for Rolfe and the sincerity of her conversion, but the marriage was without question political. Both Wahunsunacock and the colonial governor approved of the wedding immediately upon hearing of it, and a fragile peace ensued between the two peoples. Present to give away the bride at the wedding was her uncle, Opecancanough.

The marriage also proved a public relations coup for the Virginia Company. Pocahontas and Rolfe traveled to London in 1616 where they were received at the royal court and created an enormous social stir. Pocahontas's cheerful outgoingness, her newfound Christianity, her apparent fluency in English, and her "royal blood" made her the perfect Powhatan ambassador to London society. For a very brief moment it seemed that the new Jerusalem might take hold in Virginia. But Pocahontas, like almost all native Americans, had no resistence to European diseases. In 1617, a few days before her return voyage to America, she died at Gravesend.

The god most easily understood by the Powhatans was not the Christian God but the god of gunpowder. This god they learned to worship. Powhatans openly admitting that the white man's god had to be more powerful than their own, and there are cases of Powhatans begging the colonists to pray for rain to this all-powerful deity. An example of how firearms affected political, religious, and commercial relations between Indians and Europeans is an encounter in January 1609 between John Smith and Opecancanough. Opecancanough was then chief of an Indian village at the mouth of the Pamunkee River and next in line to his brother, Wahunsunacock, to be chief of the Powhatan confederation. The Pamunkee flows into the York River across the peninsula from the fort at Jamestown. Opecancanough knew Smith from the time of Smith's captivity under the Powhatans and had received, entertained, and traded with him before. Relations were often cordial, occasionally hostile, and always tense. In this instance, Smith needed food.

The Indians were increasingly suspicious of the white man's true intentions. Realizing that it was their own corn that was keeping Englishmen alive in their midst, they no longer wished to trade. Smith sailed up the York River and landed at the village with 16 armed men, where they were fed and entertained for several days by Opecancanough. But when the day for trading came, they found the Pamunkey village abandoned. Soon Opecancanough appeared with a few followers armed with bows and arrows, but very little to trade. (Recent research has shown that the area was in the middle of a prolonged draught, and little food had been harvested.) Smith rebuked Opecancanough for not following through with a previous promise to trade. The Indians claimed that few would come forth with baskets of corn for fear of the white man's weapons. In negotiations with Smith he and his brother had repeatedly asked Smith and his men to leave their guns behind on the ship. Smith, fearing ambush, always replied that putting down his weapons is something only his enemies would ask, and as they were friends, there was no need to do so. While speaking with Opecancanough, Smith received a report that 600 to 700 Indians were gathering outside the village. As his men began to panic, Smith addressed them. First (according to his own writings), he noted that he was under strict orders from London to stay on good terms with the Indians, and feared that if there were violence the Indians would appear "…saints, and me an oppressor." Avoiding a fight was essential:

"If we should each kill our man… then shall we get no more than the bodies that are slaine, and then starve for victuall. As for their fury, it is the least danger. For well you know, being alone assaulted with 2 or 300 of them, I made them compound to save my life; and we are now 16 and they but 700 at the most; and assure your selves God wil so asist us, that if you dare but to stand to discharge your peeces (firearms), the very smoake will bee sufficient to affright them."

Thus reassuring his men that they had nothing to fear if a fight did break out, Smith challenged Opecancanough to a one-on-one duel. The chief, then in his sixties and Smith being in his twenties, responded with the offer of a present for Smith. But Smith would have to leave the protection of his companions to collect it. Suspecting treachery, Smith grabbed Opecancanough by the arm and, holding a pistol to his chest, addressed the assembly (again, according to his own account):

"I see, you Pamaunkies, the great desire you have to cut my throat, and my long suffering your injuries have imboldened you to this presumption. The cause I have forborne your insolencies is the promise I made you, before the God I serve, to be your friend, till you give me just cause to bee your enimie. If I keepe this vow, my God will keepe mee; you cannot hurt me: if I breake it, he will destroie me. But if you shoot but one arrow to shed one drop of blood of any of my men, or steale the least of these beades or copper I spurne before me with my foot; you shall see, I wil not cease revenge, if once I begin, so long as I can heare where to find one of your nation that will not deny the name of Pamaunke."[44]

And so the trading began.

Opecancanough would not forget this humiliation before the eyes of his people. It was he that organized the great massacre of 1622. In 1644, at the age of 100 years, he set his people off on another desperate attack that killed an additional 500 Englishmen. He was captured and brought to the fort at Jamestown, where he was shot in the back by a soldier assigned to guard him.

The ideal of converting native Americans to Christianity was not unique to the English, nor was it new to the Chesapeake. In 1561, forty-six years before the English came to

Jamestown, two Spanish ships entered the Bahia de Santa Maria (Chesapeake Bay) and dropped anchor somewhere near Paspahegh, an Indian village on the island that later became Jamestown. Several Indians boarded, including a chief and his son. The young man so impressed the commander, Pedro Menendez de Aviles, that permission was asked to bring him to meet the King of Spain. Menendez promised to return the boy with much wealth and presents, and his father agreed on the venture. The ships set sail and crossed to Spain. After landing at Cadiz, the young American met King Phillip II and was placed with Dominican friars at Seville to learn Christianity and the Spanish language. He was baptized and, in accordance with the promise, returned to the new world with Menendez in 1563, by way of Mexico. But Menendez, assuming he would be brought back to his people, left him in Mexico under the care of a Don Luis de Velasco who became so enamored of the young man that he gave him his name "Don Luis." Fearing that he would return to his savage ways if allowed to return to North America, Velasco would not give him up. After three years passed, Menendez, still intending to fulfill his promise, organized an expedition to Bahia Santa Maria to return the young Indian. Two Dominicans friars went on the expedition to begin preaching to the native peoples, using Don Luis as an example of a worthy Christian Indian. But their ship missed the mouth of the bay, was blown offshore in a storm, and ended up back in Spain. Don Luis, now 23 years old, spent two more years in Europe, begging to be brought home for the purpose of "converting his parents, relatives, and countrymen to the faith of Jesus Christ and baptizing them and making them Christians as he was." Menendez reappeared and offered to supply a ship to take him back, along with a half dozen Jesuits to begin the task of conversion. After another two years in Cuba, Don Luis and the Jesuits finally arrived in Bahia Santa Maria. It was September, 1570, more than nine years after he had left.

Don Luis's Indian family received him warmly, thinking he had died and been reborn. The father had died and another son was ruling the country. Don Luis reportedly turned down an offer to rule in his brother's place, asserting that he wanted not "earthly things but to teach them the way to heaven which lay in instruction in the religion of Christ Our Lord." The natives "heard this with little pleasure," reported the reverend chronicler. Don Luis was willing at first to fulfill his role as guide and interpreter for the Jesuits, but soon showed signs of reverting to his native ways, and finally renounced Christianity altogether. At first he lived at the Jesuit mission which was located a few miles overland from what was later to become Jamestown, but then went to live with his brothers at his native village on the Pamunkey River. In 1571 the Jesuit mission was wiped out by the Indians, apparently under the leadership of Don Luis. Don Luis, however, had changed his name to "Opecancanough,"[45] which, in the Algonquian language means "He whose soul is white."[46]

The attempt to erase the cultural memory of American native tribes and replace it with a European spiritual heritage is best summarized by a letter home from Virginia by Jonas Stockden: "I confesse you say well to have them converted by faire meanes, but they scorne to acknowledge it; as for the gifts bestowed on them they devoure them, and so they would the givers if they could: and though they (Christian settlers) have endeavored by all the meanes they could by kindnesse to convert them, they finde nothing from them but derision and ridiculous answers...and till their Priests and Ancients (Chiefs and elders) have their throats cut, there is no hope to bring them to conversion."[47]

Jamestown was a disaster. It is a story is of exploitation, mismanagement, disease, famine, extortion, slavery, violence, and genocide, with but a moment of romance between an English tobacco farmer and an Indian princess. By all rights, Jamestown should have disappeared from the map in the manner of other English colonies in Roanoke, Popham, and Newfoundland. What kept it alive was not its economy, its strategic location, or its crusading spirit, but the constant stream of people arriving on its shore. No matter what the rate of failure, enough people would grab a toehold to keep the colony alive until it became us. It is not surprising that we, as Americans, sidestep the dismal Jamestown story in favor of another, more heroic story that began up the coast 13 years later.

When John Smith envisioned his "great cittie" on the banks of the James River four hundred years ago, he could not have envisioned the Lincoln Memorial, the New York City subway system, the fruited plains of Kansas and Iowa, or the Golden Gate Bridge of San Francisco. A defensible outpost and a steady food supply were all he could reasonably hope for, and even this was years in the making. Had he specified his dreams for Jamestown and America four hundred years into the future, they would be laughable now. He may have fancied a transcontinental horse trail, complete with water troughs, way stations, forts, castles, and ferry crossings, linking farms and villages coast to coast (if he knew where the other coast might be). We are glad that he was not overly diligent in predicting what lay four hundred years ahead for his "great cittie." It is with some risk of ridicule, therefore, four hundred years hence, that we peer into the future now. We cannot help but be wrong in specifics, so we will stick as close as possible to generalities. Mr. Smith was right, after all, in his vision of great "cities," though wrong in seeing one on a swampy island in the tidewaters of the Chesapeake.

Unlike Smith, I do not believe that the urge to expand to new continents or new worlds is of itself a good thing. Nor do I think it a bad thing. I believe it is a force inherent in life, and that humans will not stop or start it. It is not ours to will or possess, though we may be possessed by it. Life, I believe, will ride the human vehicle beyond current gravitational limitations, no matter what we may be thinking at the time. We will expound reasons for staying on Earth and reasons for venturing forth, but they will not matter. We will go. We will go as plants and animals, air and water, earth and energy, microbe and human, and there will be new Jamestowns to settle, hopefully more carefully planned and organized. We will go as people and we will go as habitat for people. We will go not as a form of life but as life formed. Humans will be in charge; we will build the arc and set the course, but the urge to go will lie more deeply within us than thought or reason. It will be the urge of rock, ocean, and soil.

Fanciful notions of space exploration are tempered by the hard reality that we have already done it. We have been beyond the Earth's pull and it is not so far. But what about staying there? It can be done, without a doubt, but why will we want to do it? The moon is a barren rock, without life, air, or water that we know of in any useful quantity. Its lack of atmosphere means not only that oxygen must be supplied artificially, but that pressurized space suits must be worn out of doors at all times to keep body fluids from exploding. It is a harsh environment. Permanent pressurized structures on the moon will provide an atmosphere suitable for terrestrial life forms, but the lack of a natural atmosphere leaves people, plants and other life forms vulnerable to meteors, cosmic rays, ultraviolet radiation, and solar flares. Buildings will have to be largely underground and time spent on the surface limited. Another, perhaps greater limitation for permanent habitation on the moon is its four-week day-night cycle. The same side of the moon faces the Earth at all times, so that as it moves in its orbit around the Earth the surface facing the sun changes. But this "day" changes only once a month. If you are standing on the surface of the moon, the Earth mostly stays still in the sky (though it will be seen rotating in place). The sun takes two weeks to move across the sky, then sets and disappears for another two weeks before rising.. This will be hard for humans and other animals to put up with, though they will adapt in time. For plants, it will be an impossiblity. Plants will be necessary for food and gas exchange within any permanent enclosure. They will have to be grown underground with artificial lighting or on the surface with orbiting solar reflectors providing nightime lighting. In time, new varieties of plants may evolve adapted to the lunar day cycle.

The moon's great advantage is its low-gravity environment. Everything on the moon weighs about one-sixth what it weighs on Earth. New compounds, crystals, and pharmaceuticals that cannot be developed at 1g (Earth gravity) or 0g (orbit or interstellar space) may find 1/6g suitable. As construction and manufacturing needs develop in space, materials that can be mined from the moon's surface will require far less energy to be lifted into orbit that those originating on Earth. Space vehicles on interplanetary missions (particularly orbital missions) departing from and arriving on the moon will require a small fraction of the thrust required for escaping the Earth's gravity, and use smaller engines and less fuel. The great near-term use of the moon's low gravity environment will be for a large visual observatory. Telescope mirrors are large and heavy, and gravity has become a limiting factor for their construction. They cannot be built much larger on Earth than they have already been built, but extremely large mirrors could be built in the lower gravity of the moon. Earth-based astronomy is also limited by weather conditions and atmospheric distortion, neither of which exist on the moon. The moon's extreme seismic stability (its lack of "moonquakes") provides the possibility of a large array of telescopes distributed across the lunar surface with their visual data combined to create a single large image. A lunar-sized array will resolve clear visual images of planets orbiting nearby stars. On the far side of the moon (the surface always facing away from the Earth), radio telescope arrays will be built, safely shielded from Earth-based radio noise.

Mars, though it has not been visited by earthly life, is far more suitable than the moon for permanent settlement. It has sufficient gravity (about 3/8g) to hold an atmosphere and seems to have large quantities of water, probably under its surface. Its year is less than two Earth years, and terrestrial life will find it easy to adapt to the twenty-five hour Martian day. (Many humans wish they could have it on Earth). Large, glazed bio-enclosures will be built and occupied by plants, animals, air, soil, bacteria, and people. The martian atmosphere is only about 1 percent the pressure of the Earth's atmosphere, so space suits will have to be worn outside, and surface exposure to cosmic radiation will likely have to be limited. In time, however, Mars may prove a viable candidate for what is known as "terraforming," or the artificial creation of an Earthlike environment for the entire planet. Mars likely had a warm and humid climate at one time, and there is an enormous reservoir of frozen water and carbon dioxide in Martian soil left over from that time. If an atmospheric warming process could be triggered, say by the artificial introduction of greenhouse gases, rising temperatures would cause water vapor and carbon dioxide to volitalize into the atmosphere from the soil, initiating further greenhouse warming. Over a long enough time, sufficient temperature increases would allow plants to grow on Martian soil outside of enclosures, and would also allow sufficient atmospheric pressure develop to dispense with space suits. As plants release oxygen into the atmosphere, people and animals would someday even breath freely out of doors. But because Mars does not have the gravitational strength to hold enough air to develop a terrestrial atmospheric pressure, the portion of oxygen in its atmosphere will have to be kept artificially high (as high as 60 percent, as opposed to Earth's 20 percent) to support free-roaming animal life.[48]

Mars's main disadvantage over the moon is its distance. Current un-manned missions from Earth to Mars are launched only when the two planets are properly aligned, which is less often than once a year. A one way voyage takes eight and a half months because the space vehicle, once it escapes Earth's gravity, falls into solar orbit and does not reach the Red Planet until it is halfway around its orbit on the other side of the sun. But this is the only the least energy route: travel time can be reduced and the launch window widened with higher energy trajectories. Energy expenditure is a limiting factor to any sort of human-led activity in the current age, and is likely to remain so throughout early stages of Mars colonization, but it may not remain so. With higher expenditures of energy, space vehicles, manned and unmanned, will be accelerated on more direct routes, and arrive on Mars in a matter of days.

Mars is the outermost of the rocky terrestrial planets, and the last with a hard outer surface (but for Pluto, which has recently lost its status as a planet). Mars is half again Earth's distance from the sun, so that sunlight falling on its surface is less than half the intensity of that falling on Earth. With the possible exception of the asteroid belt, Mars is the likely outer limit for practical use of solar energy, as we now understand it, both for plant photosynthesis and for human energy needs. Beyond Mars, the gaseous jovian planets, Jupiter, Saturn, Uranus, and Neptune, have too much gravity and too much atmosphere to serve as a base for a life colony that we can in any way foresee. But the jovian planets do have rocky moons that could be colonized, some of them with enough gravity to hold atmospheres. Jupiter's moons are currently off the list, as they orbit within dangerous radiation belts surrounding the planet, but Titan, a satelite of Saturn, is a candidate for settlement, perhaps even for terraforming. Two Uranian moons, Titania and Oberon, also look interesting, as does Triton, a moon of Neptune. The jovian planets themselves may become vast energy sources one day, as they possess isotopic elements suitable for use in fusion reactors.

Another strong candidate for terraforming is Earth's sister planet, Venus. It is 95 percent the size (diameter) of the Earth, and, with 82 per cent of Earth's mass, its gravity is only slightly less. It is three-quarters the Earth's distance from the sun, and its year is 225 days long. It is closer to Earth than any other planet. But there are two major problems with Venus: her climate and her rotation. The first is the more serious, but perhaps the more solvable. Though nearly twice as far from the sun as Mercury, Venus is hotter. This is due to thick clouds of carbon dioxide that surround the planet, trapping solar energy in a run-away greenhouse effect. Any life form descending through the clouds to the surface would be broiled in seconds. Some visionaries have suggested that Venus could be terraformed by seeding the upper atmosphere with photosynthesizing microbes that consume carbon dioxide and gradually release oxygen. As the climate cooled, plants could be grown on the surface. In time (a lot of time), the Venusian atmosphere would simulate that of the Earth. But even with an Earth-like atmosphere, Venus rotates much more slowly than the Earth, and in the opposite direction. Venusian days and nights are each 127 Earth days long, far too long a cycle for any known Earth plants. No perennial could survive a long Venusian night, though agricultural crops (most of them annuals) could grow and bear fruit in a single Venusian day, with periodic shading to simulate terrestrial night.

Some futurists claim that the best way to colonize space is to avoid planets and moons altogether. Rather than adapting to the harsh conditions of any non-terrestrial surface, it may be simpler to build large self-contained bio-enclosures in orbit around the Earth, the moon, the sun, or any nearby planet. Total control of living conditions could be achieved, and transportation between orbiting enclosures would be accomplished without the energy and thrust requirements of entering and leaving gravitational fields. To avoid the long term health effects of a 0g environment, disc-shaped colonies will create artificial gravity by rotating at velocities determined by their overall size. As centrifugal effects increase with distance from the axis, the larger the disc, the slower the rotation. Houses, paths, farms, forrests, etc. will be located on the "ground" at the perimeter of the disc, where artificial gravity is greatest. As one moves "up" towards the center of the disc, gravity decreases until, at the center, there is none at all.

Permanently inhabited or not, large orbiting structures are likely in the near future. Solar collectors, severely limited on the Earth's surface by clouds, nightime, and a constantly changing angle of solar incidence, will be constructed in fixed orbital positions around the Earth and the sun. Energy will be transmitted to the Earth by microwave. The sun emits trillions of times as much energy as humans need, most of it radiating into empty space. Of the tiny portion that hits the Earth, most is reflected by clouds and oceans or sent back into space in the form of infrared radiation. Of that which is absorbed by photosynthesis, a tiny portion has been stored underground in the form of coal, petroleum, and natural gas. The infinitessimal fraction of the sun's energy stored in these "fossil fuels" keeps human civilization alive today, but cannot be used much longer. Large orbiters that collect solar energy directly present the greatest engineering challenge of the twenty-first century. Humanity is unlikely to survive without them.

The "yang" of extraterrestrial colonization is about penetrating the unknown and extending the frontier, about propulsion, rocket fuel, orbiters, shuttles and thrusters, about artificial life support and one-way consumption. It is about setting foot in strange places and coming back before supplies run out. It is open and linear, respiration without photosynthesis, animal without plant, oxygen without carbon dioxide, humanity without biology. It is bold, extroverted, aggressive, and unsustainable. The Apollo lunar landings of the late 1960s and early 1970s are the beginnings of the yang. In time, the "yin" of nurture and plant life will balance the yang. Life within permanent space enclosures will be a circle and not a line. Colonies will be occupied on an ongoing basis by select portions of the living world, including a fair representation of humanity. Self sustaining space capsules, bio-domes, and terraformed moons and planets will be the yin of life expansion into space.

The yin need not wait for the yang. We need not be thrust into space to learn the balance of life within self-contained units. We already have greenhouses, and we already have degrees of self-containment in our houses in the form of air conditioning and central heating. Even a roof is a form of bio-enslosure. We should experiment with

greater degrees of self-containment by building enclosures
on Earth in which people, animals, plants, soil, water,
bacteria, and sunlight interact without external support.
How do you keep the air breathable for both plants and
animals? How do you prevent plant disease? Do you vary
temperatures, daylight, humidity? What are the psycho-
logical needs of people and animals in an enclosed space?
To simulate lunar or martian conditions, enclosures should
be built in deserts, over open ocean, in polar regions, and
in low Earth orbit. If temperatures within greenhouses are
kept high enough, are there plants that will grow in the
constant, low-intensity sunlight of a six-month arctic day?

The yin of extraterrestrial life is, of course, the yin of
terrestrial life. Experimentation with closed biosystems
will help people understand the Earth as a closed biosys-
tem, and help balance the yang of conquest, exploitation,
and linear consumption. Only when we know who we are
will we be ready for space.

Interstellar and intergallectic travel, as opposed to interplanetary travel, requires a new understanding of what it
means to "go" somewhere. To "go" means to change one's
position in space. But space has meaning only in relation to
objects in space. Because there are no fixed positions in the
Universe, one "goes" in relation to one's surroundings. To
go across the room or across the continent means to change
one's perceptual perspective. What is seen, heard, touched,
etc. from where one has been will be seen, heard, etc. "in a
new light," and new objects will be seen for the first time.

Travel also changes who one is with. To go to the next room, or the next town, is to be with new people, or to "go back" to people one has been with before. Very little of this will be experienced in interstellar travel.

Interplanetary travel is in relation to the sun and the planets. (Planets move in *space*, but their fixed orbits around the sun define fixed positions in *space-time*: we know where they will be when, and may therefore be said to move in relation to them.) Looking through the window of a spacecraft moving from the moon to Jupiter, one would see the sun, the Earth, Mars, and the occasional asteroid, and one would sense gradual changes in their positions. This would be, I think, the great fun of interplanetary voyaging. But looking out the window of an interstellar spacecraft moving from our solar system to Alpha Centauri (the closest star), one would notice no positional changes at all. There would be no sense of motion. The only noticable difference in the sky would be a gradual brightening, over the years, of a star in the Southern Cross, and the gradual dimming of our own star, in Casseopia.

Until arrival at a new star system, one's visual perspective would change very little. Even after arrival, one's perspective in the non-visual realms would change not at all. The olfactory and auditory realms require an atmospheric medium, likely to be found only in the enclosure one has been in throughout the entire trip. New foods are unlikely to be tasted in the vicinity of the new star, and new objects unlikely to be touched, except perhaps through the heavy protective wear of a space suit. The only real sense of "being there" will be in the visual realm, and could be as easily accomplished with a video camera. Sending *people* on interstellar missions is an awful lot of trouble and expense for what we will get out of it. And yet, again, we will go. We will not go to make money or because life is better there; we will go because that is what life does.

But it is doubtful that the "we" that goes into space will be any more recognizable than the "going." Sending whole, conscious human bodies with food, fuel, and life support across lightyears of empty space is an absurdly expensive proposition. Keeping them sane is several times more expensive. Even with the money and the technology, how can we justify putting people into an enclosure they would have no way of leaving for decades and probably for generations? There will be volunteers, of course, but how can we subject their unborn offspring to a life without blue sky and sunsets? Suspended animation of most or all of the crew would reduce food, life suppport, and recreational requirements, but there are questions as to whether it can be done sucessfully. It would substantially increase technical requirements of the mission.

Presumably, we will want to stay in touch with these people. The "we" that sends people on interstellar missions will consist of those who make the trip, those who stay home, and some sort of communication between the two. But communication will not be what we are used to. Time delays between messages will skew two-way communication beyond recognition. Even within the solar system, radio waves take several minutes to reach Mars, and an hour or so for Jupiter and beyond. Interstellar time delays will be years, decades, and centuries. A message to the nearest star system, Alpha Centauri, will take 4.3 years to arrive, and a reply will not come back for 8.6 years. Messages sent to other star systems in our local corner of the galaxy, even those visible to the naked eye, will be in transit a hundred years or so, each way. Communications between star systems will be one way, for all intents and purposes. There will be news, but no conversation. The concept of simultaneity for human activity outside the solar system will all but vanish. Reading the latest bulletin from a distant star system will be like hearing the outcome of the Spanish-American War. What sort of "we" is this? "We" is generally those of us who can *act together*. We are a "we" if we do things as a family, a community, a nation, or a world, and the strength and coherence of the "we" varies directly with the strength and coherence of our actions. A "we" without two-way communication does nothing at all.

Travel between far flung outposts will be many times slower than communication. Human led bioenclosures that manage to survive in other star syatems will each adapt to local conditions of climate, atmosphere, gravity, day length, geology, etc., and will evolve in isolation from each other. New varieties of plants, humans, animals, and bacteria will emerge under varying conditions, and new species will naturally emerge from isolated breeding populations. From the standpoint of genome diversity, this will be a good thing, particularly for humans. The genus *homo* has placed all its genetic eggs in one basket by narrowing, since the extinction of australopithicines and neaderthals, to a single species, *homo sapiens*. "Successful" animal species are those that diversify and generate new species; those that do not diversify tend toward exitinction. In the very long run of things, "we" will survive only when we are something else, or at least when we broaden our understanding of what "we" are. Interstellar and intergallectic space travel provides the hope for very long term survival of something like ourselves, though by our current identity limitations, there is still nothing in it for "us."

Still, we will go, in a very different sort of going. We will not go in our bodies and there will be no coming back. We will go as genetic material or just genetic information, of our human selves and our plant and animal selves – with instructions. Information does not weigh much, and can be accelerated to near light speeds with little rocket fuel, or none. Additional information will be sent from time to time but, like a dandelion casting its seed to the wind, we will never hear back. We will not know what happens. We cannot know what conditions will be, so we will emit a steady stream of something that is ourselves in as many directions as we can imagine, hoping some of it will land on fertile soil and take shape. It has all been done before.

Higher Realms
of Consciousness

This chapter will be as difficult as chapter IV. But if you waded, skimmed, or swam through that one, you will survive this one, too. I have avoided physics, for the most part, but cannot avoid an analysis of conscious experience that is unusual and perhaps a bit uncomfortable for many readers. We are not used to looking inward and trying to describe what we see in analytical terms, and it is often the terms themselves that get in the way. *Perception, sensation, conception, observation, composite structure, realm, coordinate, doing,* etc. have meanings that I do not intend here, and there is bound to be confusion. But with patience, the picture should become clear.

The whole of consciousness is divided (with a degree of arbitrary subjectivity) into perception, observation, and conception. *Perception* is the five sensory realms, corresponding to quantifiable dimensions of space and time. It is direct experience. *Observation* is a sixth dimensional realm that corresponds to the non-quantifiable dimension of *order*. Observation is *indirect* experience: what somebody tells you they see or hear, through language or electrronic imagery. The *conceptual realms* are non-dimensional – they are not experienced in space or time – though one of them, the *practical realm*, "coordinates" with space-time. Objects that you conceive in this realm become dimensional (you can see and touch them in space-time) when you *do* things.

My purpose in writing this book is to suggest that it is human consciousness rather than the human body that is evolving in the current era. Consciousness is evolving so rapidly right now that, to continue to live successfully, we must recognize what is happening to us. We must be more conscious of consciousness. We must see what is going on and adapt. Electronic technology, external to the human body, is weaving us into a higher form of biological presence on earth that must be understood, I believe, from the standpoint less of objective reality, than of direct experience. We must look "inward" and become more familiar with what is going on in the mind and how that relates to the "outer" world. The outline I suggest in this chapter, and in the book as a whole, is by no means a final word on the subject; it is, rather, an opening word that I hope will encourage thinking people to explore greater realities now available to human experience.

THE OBSERVATIONAL REALM

The composite structure of consciousness (its divisibility into realms) explains the relation between cellular and organic experience. Taste and touch, the chemotactile realms, are directly cellular, where smelling, hearing, and seeing are experiences of the organism as a whole, but *reducible* to the chemotactile experiences of sensory cells. I hope to show later that the composite structure of consciousness also explains the relation between individual and social consciousness. *Observational* ("objective") experience is that of society as a whole, but reducible to the sounds and symbols of individual hearing and seeing.

Before moving to the second of these considerations, it is important to distinguish between consciousness and *self*. There is a strong tendency to identify these, particularly in western traditions, but they are not the same. For purposes here, *consciousness* means seeing, hearing, touching, smelling, and tasting, as well as thinking, feeling, imagining, etc., but it does not mean perception *of* something *out there* waiting to be perceived. It means experience – not a physical process that depends on a reality external to experience. Consciousness is the screen – not something on the screen. In this way of understanding there is no material substance and no physical object independent of perception. What you see is what is. It is the coordination of sensory potentials that seems to indicate *matter* – where and when you see an object is also where and when you can touch it, lending the object a sense of independent "material" existence. With no object independent of perception, there is no subject independent of perception. Consciousness is the *seeing*, not the *seen* or the *seer*. It is in this sense purely passive, and best understood without reference to self or to *doing*. Self is a focus of order that becomes necessary for *doing*, but it is not necessary for consciousness. Consciousness can exist without self, though the self, naturally, cannot experience it as such.

The existence of dimensions within consciousness does not mean that the physical world is *in me*, or *in you*. Nor does it mean that the physical world is *my idea*, that it *doesn't really exist*, or that it can be *manipulated at will*. These are all projections of self. What it does mean is that the physical world exists within the larger experience of being. It is the box that we see in the form of a space-time screen.

In the structure of perceptual consciousness we have seen how one realm, the visual, is *reducible* to another, the tactile. Vision can be reduced to *nothing but* the tactile experience of retinal receptor cells. This is the key to understanding how one realm is built from another. The auditory realm is reducible to the tactile experience of cells in the tympanum (ear drum), and the olfactory realm to the chemical experience of cells of olfactory organs. The tactile realm is in turn reducible to the chemical experience of cells, in that pressure on cell membranes produces movement of potassium ions on its surface. In each case it is the *order* in the information received that produces the new realm of perception.

Order creates wholeness over and above parts. The orderly arrangement of minute tactile sensations into patterns produces the richness and complexity that makes them visual and not tactile. Reduction ignores order. It favors parts at the expense of the whole to the point where the whole disappears altogether. It is an extremely limited view on its own; but combined with a holistic view, it reveals the relation between parts and whole. Along with holism, reduction shows the structural relation between cellular and organic consciousness.

The relation between perceptual and observational consciousness is also revealed through reduction. Observational consciousness is indirect perceptual experience, or what you know of the world through observers, or other people. You may never see the Great Wall of China, the far side of the moon, or your friend's house on the other side of town, but you are aware of them through language, television, radio, telephone, newspapers, photography, etc. Observational experience is reducible to perceptual experience in that you *hear* sounds as your friend speaks and *see* ink arranged into words in the newspaper, but your experience is much more than what you hear or see directly. What you actually hear or see is the sound or sight of the words. It is the order of the sounds coming from your friend's mouth and of the ink on the page that creates a realm of experience greater than the sum of its parts. This is the larger world that you do not see or touch directly but that you experience as a communicating human being. Holistically, it is more than the sound of words or patterns of ink on a page, but it is also a greater world in the sense that it is more and bigger than you will ever perceive directly. Your see with your eyes the sun coming up in the morning and going down at night, but you hear from others that it is the rotation of the earth that you are seeing. Through observational consciousness you are aware of a world infinitely larger than you can see with your eyes.

The observational realm is coordinated with the perceptual realms as a whole the way they are coordinated with each other. The information potentials for hearing, seeing, touching etc. are inter-coordinated dimensions of space and time: Where and when you hear an object in space-time is where and when you will see it if you are looking or touch it if you reach out toward it. Observation is coordinated the same way: If an observer tells you there is an object at a certain point or range of points in space-time, you will potentially hear, see, or touch it there. This means that it is a physical object, subject to scientific analysis. If he says it is there and you cannot perceive it, he is hallucinating or lying, and what he says is not observational information. The scientific method is the process by which perceive reality becomes physical reality.

The question naturally arises as to whether an observer is *actually seeing* what he is describing in words. If he says, for instance, that he sees a red house on the street corner, is he really seeing it the way you see it, and is his concept of "red" the same as yours? Could he be seeing what you see as blue, simply using the word "red" to describe it? You turn and see the house there, so there is no question as to the observational validity of his words, but all you *actually experience* from him is the sound of his words and their particular order. They align dimensionally with what you perceive, but do they prove he is actually conscious? There is, of course, no way to know. The question may seem a rhetorical game, but it clearly highlights the problem of understanding consciousness *inside* a living being.

If you think for a minute you might catch a glimpse of what I mean by the structure of consciousness: Listen sometime to what someone is saying and ask yourself: "Where does the consciousness in these words come from?" Normally you assume that it is some mysterious fluid churning about inside his head, but what do you actually experience? Look more closely!

Observational consciousness is *in the experience* of hearing his words – not something in him or in you. Being is in the words, said and heard; his being is not separate from your own. The importance in hearing his words is not whether or not there is consciousness in him; the importance is that by hearing his words you see through his eyes. You do not have to look at the house to know it is there, and that it is red. There is a realm of consciousness here greater than direct perception.

You create observational consciousness when you encode perceptual experience into words and symbols. When you are speaking with other people, you see *the same thing* as they do only in that what you see is dimensionally coordinated with what they say they see. Observation in human societies is in the form of spoken or written language, graphic drawing, and electronic media. It encodes the dimensional coordinates of what an observer perceives in a manner that is perceivable by any other observer under the same conditions. It is defined, therefore, as *potential perception*. It is what anyone else would see. The words and symbols that you hear or see are perceptual consciousness, but so highly ordered that they create an entirely new realm of consciousness. Observation is reading, looking at pictures, watching television, surfing the net, or listening to the radio. It is experiencing things you would see directly if you were there. It is the life of society as a whole. It is not in individual people any more than seeing or hearing is in individual cells.

Observation is not limited to human societies. As we have seen, some insect societies are so highly organized that there has to be a wealth of information flowing between individuals and probably also a degree of dimensional coordination between that information and individual perception. There is likely some structure of observational consciousness in these cases. It seems that the phylum just below ours touched on observational consciousness a hundred million years ago but only got so far with it. There can be little doubt that observational consciousness is far more highly developed in vertebrates, and humans in particular. It continues to evolve among humans at an extremely rapid rate. A generation or two ago human life consisted mostly of what individual people saw and heard directly; now human life consists largely of what is seen and heard through electronic media. There is less perceptual consciousness and more observational consciousness in what it means to be human.

Scientific investigation is the systematic growth of the observational realm. The scientific method converts perceptual experience of the scientist to observational experience of the scientific community and the rest of society. A scientist tests what he sees to be sure that it can be seen not only by himself but by anyone at any time under the same conditions. He converts perceptual experiences into symbols and communicates them through a medium, where they become subject to critical review. Experiments must be repeatable and the same results observed by others. Interpretations may vary as to the meaning of data, but to be science, the data must be incontrovertible. Science restricts itself to that portion of conscious experience that can be measured, encoded symbolically, and potentially experienced by others, or in other words, to that which is in the box. It cannot concern itself with such things as a "field of vision," the "structure of consciousness," or the "experience" of a cell or insect society. These are outside the observational realm and beyond the bounds of science. They are as real as science, however, and as important to understand.

The observational realm has grown so rapidly and become so powerful a presence that it is often taken for reality itself. Imagination, direct perception, and spiritual experience are trivialized or dismissed because they cannot be measured and experienced directly by others. Life is reduced to physiological function. This is a mistake – experience is much broader than that which can be packaged and perceived by others. We all dream, imagine, perceive, conceive, believe, and have being that is not available to science. These are all true and valid aspects of being, and life is better understood when science understands itself to be a limited realm in which collective consciousness is structured from individual experience. At this moment in evolutionary history, science is the moving force of life, but life will evolve successfully only if science stays within the bounds it has set for itself.

Perception and observation are *dimensional* realms of consciousness. Images within them are perceived and observed as objects in space and time. But not all consciousness is dimensional and not all images are objects. Thoughts, dreams, ideas, images and hallucinations are, for the most part, less vivid and enduring than perceptions and observations, but they are distinct from perceptions and observations only in that they are non-dimensional. They are not experienced in physical space-time. (They may be in *time*, but not in *space-time*.) Non-dimensional images are as real as dimensional objects, the only substantial difference being their context. A physical object is an image in a dimension.

Non-dimensional images are experienced in *conceptual consciousness*. Conceptual consciousness is mostly unstructured and often chaotic. Like the perceptual and observational realms of dimensional consciousness, it consists of separate realms, but realms not as distinct or as easily delineated. Causal relations that operate within dimensional realms do not penetrate far into conceptual consciousness, and it is a mistake to assume that physical objects *cause* conceptual images. Conceptual images and physical objects are of the same primal substance; the difference being that a physical object is a set of interrelated visual, tactile, auditory, chemical, tactile, and observational images.

I use the term *conceptual consciousness* in so broad a sense here that it may cause confusion. I mean by it any form of consciousness that is not perceptual or observational: Some of its contents may not be "conceptual" in the narrower sense. Conceptual consciousness is unstructured, primal, and non-dimensional, and it is the source from which new realms of consciousness arise. An example of a conceptual realm is the realm of imagination. Another is dreaming, and still another spiritual experience. Each has distinct characteristics. Imagination is not coordinated with the perceptual realms but, unlike dreaming, it may be experienced at the same *time* as perceptual consciousness (i.e., when awake). The spiritual realm is similar in many respects to imagination or to dreaming, but distinguished from either by transcendence of self. If experience and not dimensionality is the test of "reality," all of these realms are as real as seeing or hearing.

The Practical Realm

There is a realm of conceptual consciousness related to imagination that is of particular interest in understanding the force of life. This is a partially structured realm in which physical objects are arranged not as they *are* in space-time, but as they *could be*. It is, therefore, a potential based on space-time structure. Unlike imagination, it is constricted by *doability*, or to a practical convertibility to physical reality. The furniture in a room, for instance, is experienced in this realm, not where it is seen and felt (where perceived), and not on the wall or on the ceiling (where imagined), but in some other location on the floor where it *could* be. The sofa, for instance, while perceived by the door, may be conceived near the window. The sofa near the window (the conceived sofa) is not coordinated with the physical dimensions and there are no visual, tactile, or other perceptual images of it at that location. Nor are there potential perceptual (observational) images of it there. It is not in the physical world and no observer reports seeing it. But you may wish to move it there. You may wish to bring a conceptual image into space-time by creating a new physical order that *is* seen and touched by you and reported by observers. This conceptual consciousness is what I call *doable thought*, or the *practical realm*. It is a dimension larger than the physical world in that it includes all possible arrangements of objects within the physical world, but smaller than imagination in that it is confined to the physically possible.

Doing requires coordination of the practical realm with the tactile realm (the body). You cannot *think* the sofa from the door to the window; you have to push it there. You have to coordinate your body with the arrangement you have conceived. It is this act of coordinating doable thought with the physical dimensions that constitutes *doing*. Once done, you will perceive the sofa at its new location, and your friends will say they see it there, too. All life *does* by creating order. This is the physical manifestation of life – the force of life.

The coordination of the practical with the perceptual realms is the *self*. Self is a focus of order, and it is through self that conceived order becomes physical order. The universe outside of self loses order as time passes. Stars burn out, systems wind down, moving parts give way to friction. Total energy remains the same, but the high-grade energy of discernable objects and living things gradually decays to the low-grade energy of undifferentiated mass and heat. The arrow of time points from higher order in the past to lower order in the future. But order *does* increase in local concentrations, especially on Earth. The sun bathes the Earth in undifferentiated energy that is absorbed and converted into plant tissue and animal locomotion. Water, nitrogen, and carbon dioxide are converted into orderly proteins and carbohydrates; cells divide and form embryos; big fish eat little fish; primates descend from trees and write poetry. Things get better – more evolved, complex and organized. The sun slowly burns out, but the Earth makes life from the sun's light. The universe winds down; life winds up.

Order is as fundamental as space, time, or mass. It is not in the dimensions, but one of them – at right angles. It is itself a dimension, however unquantified. It is experienced when the sound of a voice becomes an event far away.

Self is the point of view from which order looks orderly. But order is not universally orderly from every point of view. There are other selves. Little fish do not see order in big fish. Poetry is not orderly for those who do not care for it. Army ants may produce healthy new broods as they march through your house, but your focus of order will likely differ from theirs. Firebombs exploding in your city are of the highest state of order for the enemy. Order always has a focal point – the self – whether it is a group or an individual. The self is not pure seeing – not seeing what is – but seeing what could be and how it could be. Self is seeing with an attitude. It is a curvature of energy, the angle from which bigger, better, healthier, and more complex is appreciated, and done. It is how the word *good* is defined. Self is the curvature of consciousness that creates order from a particular point of view. It permeates consciousness in and out of the dimensions: You can watch it take shape in the world and in the mind.

A way to see the self is to sit quietly and take the time to watch how a path of thinking develops. Thoughts are random at first: The mind considers this and that, passing from one direction to another without overall plan. After a while thoughts begin to fall into distinct patterns: your leg hurts; you have an errand to do; you should have said this or that. As the self becomes restless it begins to curve thinking towards coordination with the body. You could move your leg, drive down Main Street and park behind the drug store, call the salesman and give him a piece of your mind. You cannot sit there forever – eventually you have to do something. Self creates the order that becomes doing.

The most highly defined self is your self. Your self is a unique perspective in space-time with the only direct access to the perceptual realms of consciousness. "Good" things are those that keep you fed, warm, empowered, entertained, and reproduced. Bad things let you become hungry, irritated, cold, weak, frustrated, or dead. You, like other selves, create order in the world to keep good things happening and bad things not happening. Perceptual consciousness is not all of your self, as you also think and do for the larger, less-defined collective selves of family, work, community, nation, congregation, or baseball team.

Collective selves, human or otherwise, are less highly defined. They are a flock of geese landing on a lake, a colony of club mosses struggling to survive in a hostile environment, or a political action committee trying to get a bill passed into law. Collective selves compete, overlap, come and go, rise and fall, each characterized by purpose and by the ability to *do*. A collective self is a collective practical realm of which you may be part. Parts are separately visible and may act at times against the interests of the whole. An individual goose, moss, or committee member may scatter, disappear, or re-form with another collective self. Without purpose or direction, a collective self loses its focus and disappears altogether. It creates new order or becomes lost to raw material for other selves, like loose polymers in the primordial soup. Collective selves have differing points of view and conflicting missions. They act selfishly. Creed and nationality are collective selves: selfless from within their point of view but selfish to outsiders. Humanity is a great self, but does not currently have the purpose or definition of national selves. It is able to do very little collectively.

Self is not the same as consciousness. It is never whole. It is always incomplete, always conceiving that which is not, and always trying to bring order to a physical world that resists. It compares what is with what could be, and is never happy. If there were no self we would be happy with what is and there would be no life.

The Equivalence of Self

The construction of observational from perceptual consciousness is accomplished through the equivalence of self. *Equivalence* does not mean equality in all respects; it means equality in one or more respects. We remain unequal in any number of ways, but from the standpoint of observational consciousness, we are the same. It does not matter who points a camera at the Great Wall of China; we all see the same image. As long as there is potential seeing, it does not matter who does the actual seeing. The equivalence of self creates a society from an organism and an organism from a cell. It is the glue that holds sponge flagellates and human civilization together.

Your self and other selves remain physically distinct but are related through structural equivalence. What you perceive is physically different from what you hear others say. When an observer speaks of a red house, you do not know what she sees or if she sees anything at all. You cannot experience her seeing a red color, hearing a bird sing, or feeling a bellyache. You hear her making sounds that lead you to believe you would experience these things in her place. But you do not actually perceive them. It is in the structure of the information that you and she are equivalent, not in the experience. There is no *other* perceptual consciousness. Nobody *else* sees, hears, smells, touches or tastes. From the standpoint of perceptual experience, every other self is empty. From the standpoint of observational experience, every self is a universe.

From the perceptual standpoint, other selves are "nothing but" order increasing in time. They are reducible to order because observation is reducible to perception. From the holistic standpoint, other selves are as alive as you. You feel their pain though you do not perceive it. They and you are equivalent because observational consciousness is a wholeness of which direct perception is a redundancy. Direct perception is not necessary; you do not have to feel the bellyache to know it is there.

The observational realm is created from nothing. There is no logic to it. At a very young age you seem the only self because you perceive nothing else directly. You feel your hunger and not mine. Later, you learn to do unto me as you would have me do unto you, though you feel directly only what is done to you and not what is done to me. Ultimately, you cannot deny the existence of order outside of yourself, so you posit, without proof, the equivalence of perceptual with observational experience. This is how you get from being an organism to being a member of society. But there is no science to it. The equivalence of self is a leap of faith, a spiritual awareness that links your self to the order around you and creates the higher self of society. It is the essence of religion and the illogic of compassion. It is an entirely separate realm of conscious being.

The equivalence of self is not visible to science because it is more fundamental than science. It is the underlying structure of science and cannot be seen within that which it is.

The equivalence of self converts a living wholeness into a building block of a higher living wholeness. It is the structural principle of multicellular and social consciousness. It is the soul of the Kingdom.

The Screen

Without mystery there is no understanding.

In straying beyond the bounds of science we have wandered into a conceptual territory that is not as precisely describable as we would like. Yet we must go there and report what we see. We have to make sense of conceptual experience even if we cannot nail it down precisely. At this point it may be helpful, therefore, to summarize what I am trying to say by describing a model of consciousness, or at least, of the dimensional portion of consciousness. It will be one of many possible understandings and carry no objective truth in the scientific sense. Its truth will lie only in its usefulness.

The fundamental unit of consciousness is the *realm*. The five sensory realms constitute *perceptual* consciousness. These realms are distinct from one another and dimensionally structured, in that information in any one of them is potentially perceivable in every other at the same coordinates in space and time. You hear or touch what you see when and where you see it. Dimensional coordination is the space-time box that gives objects of perception the semblance of independent material existence. Two of the perceptual realms, the chemical and tactile, are associated with single-celled life and with the plant and fungal kingdoms; the other three, olfactory, auditory, and visual, are associated only with the animal kingdom. Information in each of these three realms is reducible to the other two, that is, to the chemical or tactile experience of individual sensory cells. Light and sound are reducible to the tactile experience of retinal and cochlear cells, and smell is reducible to the chemical experience of olfactory cells.

Less distinct are realms of *conceptual* consciousness. These are *non-dimensional*, and with one exception, do not coordinate spatially with other realms. They include thoughts, dreams, ideas, hallucinations, hopes, fears, and spiritual experience. They are as real and valid as perceptual experience, though they cannot be independently verified with experience in other realms, as can dimensional experiences. They are not "material" or "physical," and remain a portion of consciousness outside of the screen model of dimensional consciousness discussed below. I have discussed conceptual consciousness mostly for the purpose of identifying the *practical* realm, which has a conceptual connection to perceptual consciousness.

The practical realm, or *doable thought*, is easily identified. If you sit quietly and watch what you are thinking, things will pop up right and left that you should do, will do, or should have done. Later, as you are up and about, you may notice that as you do things you are creating order that originates in this realm. You may notice the connection of this realm to the tactile realm (the body). You cannot do anything without the body – it can be as small a part of your body as your tongue and vocal cords. Doing is the body moving objects in space-time to coordinate with the practical realm. Conception becomes perception through doing. Living beings are discernible in space-time by their ability to do, which is their ability to create *order.*

Order exists always with a point of view, or *self.* Your self coordinates the practical and perceptual realms to do things; other selves are foci of order without direct experience of the perceptual realms. Selves do not contain consciousness. Consciousness without self can be experienced, but not by the self.

In addition to the conceptual and perceptual realms is the *observational* realm. It consists of experience through language, numbers, electronic media, etc. Information in this realm has to be *potential perception*, that is, it must be observable by anyone at any time under the same conditions. This is what distinguishes it from conceptual consciousness. It is created through the moral equivalence of separate selves. Observational consciousness is a higher form of life, but life cannot understand itself exclusively in observational terms.[49] This particular realm is currently in an extremely rapid state of expansion due to the growth of scientific knowledge and electronic media.

Now let us try to describe how these separate realms are assembled into a dimensional structure that we actually experience. Space is three-dimensional. A physical object in space occupies a range of points in length, width, and depth, but because objects move in space, we have come to understand time as a fourth dimension. A point in space *and* time is known as an *event*, and an object in motion through space can be understood as a range of events in four-dimensional space-time. This is "the box," the physical universe as it is commonly understood. As an operational model, it gets us through the day, but it is far too limiting for a complete understanding of consciousness. In the box, "we" are separate physical objects inside the space-time universe who happen to be "conscious." This common sense model assumes that physical reality is more fundamental than consciousness and therefore cannot explain how organic consciousness arises from separate cells or how observational consciousness arises from separate organisms.

The alternative model I suggest for the relationship between consciousness and physical reality is the *screen*. The screen is like the pixel screen of a computer or television, but it is holographic; it has three dimensions of space. It is not *in* space, it *is* space. It is not a physical object of experience; it is the framework within which we experience physical objects. It may be properly considered the physical world itself, if by "physical world" we mean the framework of perceptual experience. It is the experiential identity of subject and object. Each dimension of which the screen is structured corresponds to a realm of perceptual or observational consciousness.[50] A dimension, whether of space or of time, is an information potential for that particular realm of consciousness. A physical object is an actual perception at a location in this potential. There are six realms of perception and observation – if the screen had six-dimensions of space, or five of space and one of time, the correspondence between dimensions and realms would be plain, but the screen has only the three dimensions of space that we experience as the everyday world. How, then, can six dimensions fit into four? The answer is *foreshortening*.

A photograph of a 3D object in space is a foreshortening of one space dimension. Three dimensions are collapsed into two in order to project the image onto a flat surface.

The depth dimension does not exist on the paper of the photo, but is "seen" in the shortened dimensions of length and width dimensions. Depth values directly in the line of sight are reduced to zero, while values at right angles to the line of sight are not reduced at all. 3D is squeezed into 2D through foreshortening proportional to the angle from the line of sight. By making the photo a moving picture, we may add a time dimension. The screen of a moving picture is 3D: two dimensions of space and one of time. We can make it 4D by turning the 2D screen into a 3D hologram. This is the model of physical reality that I am suggesting. We experience the tables and chairs of ordinary living on this holographic screen. The fifth and sixth dimensions are imposed on the screen by foreshortening not in space dimensions, but in the *time* dimension.

The fifth dimension, mass, is familiar as a dimensional value, though not as a physical dimension. Physical values of force, momentum, energy, pressure, etc. are all expressed in components of space, time, and mass. Mass itself, however, is "seen" only through *acceleration*, in terms of meters per second *per second,* and it is in this second "per second" that the mass dimension is revealed foreshortened in time. (On the cosmic scale, the mass dimension is also revealed as a curvature of 4D space-time into a fifth dimension.) A physical object is a shape in five dimensions: it has extension in space, time, and mass, though its mass value is not seen at any single event in space-time. Mass is revealed only through behavior foreshortened over an extension of time.

We cannot see the extent of the mass dimension the way we can space and time; mass values are not seen as actual locations within a potential as with space dimensions, and we tend, therefore, to think of mass as substance within individual objects, rather than as a fundamental structure of the universe as a whole.

The sixth dimension, *order*, is also foreshortened in time and also seems to exist within objects rather than as a fundamental structure of the universe. Unlike mass, order is not quantifiable and is not a component of physical values such as momentum or energy. It has no distinct physical manifestation as such, but is universally recognized as an indicator of apparent consciousness "within" observers. Only those physical objects capable of observational consciousness (living beings) display orderly behavior, which may consist of building nests, stepping around obstacles, communicating through sounds or symbols, or simply moving in a purposeful manner toward or away from other objects. Non-living objects are random in this dimension. Where the fifth dimension is foreshortened in the fourth by the difference between uniform velocity and acceleration, the sixth is foreshortened in the fifth by the difference between uniform acceleration and orderly non-uniform acceleration, i.e., by a body moving in a manner that makes it look alive.

Because order is a dimension additional to space-time, and not *in* space-time, a 6D shape (living being) need not be contiguous in space. Many *are* contiguous. Living organisms, for instance, are seen as a single range of points in 4D space-time (a single object), but other living beings, like ant colonies or human societies, are many separate ranges of points in space-time (and mass) unconnected in space. There are spatial gaps within the being of the whole. The life of the colony, or of the society seems, for this reason, to be beyond the limits of physical reality, but it is not. It is only beyond the limits of four-dimensional physical reality.

This leaves room for a new vision of the force of life. Life is a shape in six dimensions. Individual organisms are visible in 4D space-time, as are inanimate objects, but their patterns of motion differ. Living objects move in ways that create the order that keeps them alive. But there are also connections among living objects that are not immediately visible in space-time, such as species-specific behavior, flocking, and societal organization. 6D shapes are flattened into 4D as an owl swoops down for its prey, a school of fish turns in unison, or a knot of traffic stops at a cross road. Order here is an invisible shape that connects living organisms the way invisible earth beneath the ocean connects the islands of an archipelago.

On the 6D screen, evolution can be understood as the constant movement of life as a whole towards order, against a background of increasing physical disorder. A new gene, a new organ, a new species, rather than a chance improvement in a random interaction of particles in space, time, and mass, becomes a deliberate motion into the separate dimension of life. In the box, the force of life looks like a random stream of ordinary patterns pushing through space, time, and mass, without purpose or direction, only occasionally tripping over a useful gene, appendage, or behavior pattern. On the screen, magic foreshortened into the orderly motion of other beings is the very magic of thought, direct perception, and being itself. Separation is the illusion of space.

Human beings at this moment in evolutionary history have a unique existential perspective. We are multi-cellular organisms evolving rapidly toward a conscious society. We experience directly the id of cellular demand for food, oxygen, homeostasis, preservation, and reproduction, and the ego of a unified self that can *do* to meet these demands. We witness in our minds the intercellular politics of competing urges that lead to action for the benefit of the whole. We are the greater self of our cells. At the same time, we see through the eyes, minds, cameras, telescopes, and particle accelerators of other people, and have learned to equate the perspectives and interests of others with those of our own perceptual sphere. We have political organizations that do collectively what cannot be done individually.

We witness, on a daily basis, the being of a highly evolved cellular organism and, at the same time, the process by which life evolves to a yet higher level of being. We are in a position to see, and to be, the whole of our cellular parts and the parts of our societal whole.

VII

———

Death on Earth

From the scientific point of view the experience of death does not exist. It cannot be verified. Death, for science, is non-life: the end of process. Death interests science only in that it shapes life the way a chisel shapes sculpture. The force of death gives form to the species, to the body, and to the chemistry of the cell. It determines which way things go by cutting off things that do not go there. It directs the course of evolution, the history of nations, and the path of walking down the street. It gives overall shape to life on Earth. Without death life would have no meaning.

Through mass extinctions death has sculpted the overall shape of life at least two-dozen times in the last 600 million years. Mass extinctions are the dying of whole species, genera, families, and orders, and of most individual organisms in the living world. The most extensive extinctions are the "big five," of 430, 380, 250, 220, and 65 million years ago. The greatest by far was the "End Permian," or P-T event 250 million years ago. Nearly every living plant and animal died and 90 to 95% of all species went extinct. The P-T ended the Paleozoic era and came close to ending life completely. It is known as the P-T because it took place at the boundary between the Permian and Triassic periods of geological history. Less severe but better known than the P-T is the more recent K-T or "End Cretaceous" event of 65 million years ago, the one that killed the dinosaurs. An even more recent extinction, not yet among the *big five,* began 10,000 years ago and continues through the present.

THE P-T EVENT

Life was pretty good for three hundred million years before the P-T event. The continents were separate, but drifting slowly together to form a single supercontinental landmass, *Pangea.* The Cambrian Explosion had seeded the seas with dozens of new life forms, while newer forms were busy invading the land. Multicellular organisms were evolving in three separate kingdoms: Animals, plants, and fungi. Life on land was harsher than in the sea – temperatures were more extreme, gravity worked against mobility and seed dispersal, and oxygen toxicity destroyed living tissue. But there were advantages to terrestrial life. Air was a better medium than water for gas exchange. Transfer and absorption of nitrogen and carbon dioxide were easier, and once animals adapted to terrestrial conditions, gaseous oxygen proved to be a faster and more efficient vehicle of metabolism. Land was a better source of potassium, phosphorus, or calcium, though water was still necessary for their absorption. Minerals were less available in the open ocean because they sank to the bottom where there were fewer plants and animals. Most marine life inhabited the upper layers of the water column where there was plenty of sunlight. Minerals became available only through *upwelling*, or the mixing of surface waters with deep ocean waters that brings nutrients to the surface. But upwelling occurred only in certain locations and under special conditions. It was an inorganic process that life depended on but could not control. On land, nutrients were conducted against gravity through *transpiration*, an organic process that life does control. Transpiration brought dissolved minerals from the soil up through the roots and stems of plants. It was far more widespread than upwelling, and promoted a diversity of photosynthetic production on land greater than in the sea.[51]

Photosynthesizing organisms in seawater were mostly single celled, and did not have to conduct water far. To live and prosper on land, plants evolved vessels in their leaves and stems and roots (*xylem and phloem*) to carry moisture and food to tissues that were not in direct contact with water. Plants may have acquired these tubular organs by co-evolving with fungi. They absorbed water and minerals through their roots by creating suction in their leaves. Evaporation through *stomata*, or small openings on the underside of their leaves, drew more water up through the plant body. Most of the absorbed water passed through the plant body in this manner and was released unchanged. A small amount was broken into hydrogen and oxygen for *photosynthesis:* The hydrogen was combined with carbon to form carbohydrates and the oxygen was released into the air. Transpiration worked well as long as there was plenty of water available; if it was not, plants were in danger of losing too much moisture through the stomata. To prevent drying out, they evolved *guard cells* on either side of each opening. Under low water conditions, these automatically closed the stomata. When the water supply improved, they become *turgid,* or swollen, automatically re-opening the stomata.

The opening and closing of stomata by guard cells in reaction to external conditions was a very animal-like function, but it was accomplished by plants without sensory organs, nerve cells, or muscle tissue. There were no specialized cells that detect moisture levels, no information processing centers, and no coordinated contractions. The plant as a whole did not "know" conditions were dry or "tell" guard cells to close stomata. Guard cells experienced conditions directly and responded only in a pre-determined manner. There were no possibilities to consider. The plant had no experience qualitatively distinct from that of its cells, no coded information, no intercellular media, and no sense of space within which to react to the environment. Guard cells just dried out as the plant dried out, and fell into a closed position around the stomata. The automatic adjustment of transpiration in reaction to local conditions was the critical step in the evolution of plant life – and thereby of terrestrial life as a whole – but occurred without the structural evolution of consciousness found in the animal kingdom.

During the *Carboniferous* era (362-290 million years ago), plants were the dominant creatures of the Earth. From tiny shoots hugging the ground near ponds and tidal pools, they quickly evolved into great forests of conifers, ferns, club mosses and horsetails. Tall canopy trees, understory trees, climbers, epiphytes, and groundcovers took hold and covered the continents. Radically different in shape and appearance from anything found in a tropical forest today, they were functionally similar to their modern counterparts, each struggling in its own way for light, water, and a means to propagate. All were spore bearers. They had no flowers, pollen, seeds, or fruit. They depended for reproduction on spewing millions of dormant spore cells into the wind. A few seed-bearing ferns evolved late in the Carboniferous, but did not survive the P-T extinction. Some club mosses reached up to 50 meters tall, taller than most trees now. Their trunks had xylem, or water-carrying vessels but no phloem, or food-carry vessels, meaning that the height achieved by these plants was not for spreading foliage in the sunlight, but for spore dispersal. This indicates how difficult it is for land plants to spread offspring over distances. They would soon seek a better way to disperse their young.

Forests began in coastal areas, floodplains, and wet swamps of the equatorial regions and gradually moved inland and upland. Intense photosynthetic production during the Carboniferous era produced more coal than any other geologic period. But Carboniferous plant life, though abundant, was far less diverse than today. There was as much total biomass, but fewer species occupied the land.

Animals were moving out of the water too, but not as fast and not far inland. Dragonflies and mayflies, two primitive flying insect orders, were hovering over ponds and lakesides, but they depended on water to lay eggs and never strayed far overland. Other arthropod species included a meter-long scorpion and *Arthropleura*, a millipede-like animal over two meters in length. Amphibians evolved from fish in shallow tidal pools, developing air-breathing lungs and legs able to support their body weight on land. But these, too, depended on water to lay eggs. Most lived at the edge between water and land, hunting insects, shrimp, and other amphibians. One amphibian, *Eryops*, was up to two meters in length and resembled the modern alligator with strong jaws and sharp teeth. It hunted fish, small reptiles and other amphibians. Early land animals were *dentrivores*, not herbivores. Rather than eating living vegetation, they browsed through leaf litter and plant debris for decaying plant matter. There were no large four-legged herbivores, and no large land carnivores. Vertebrate herbivores evolved only in the late Permian. Until then the great terrestrial forests remained free of large animals. Insects and a few small amphibians and reptiles were learning to eat soft shoots and fruiting parts of some plants, but the most abundant source of carbonaceous matter – foliage – remained unexploited. The great terrestrial symbiosis between the plant and animal kingdoms had not yet begun.

At the beginning of the Carboniferous the climate was hot. Carbon dioxide levels in the atmosphere were high. But over millions of years, plants grew so extensively that they used up most of it. Carbon dioxide went from as high as 0.5% to a low of 0.02%, one twenty-fifth as much. This led to gradual but drastic global cooling. Photosynthetic production transformed enormous quantities of atmospheric carbon to plant mass, as it does now, but there were not enough microbes and animals at the time to restore the carbon dioxide to the air. Nobody was eating enough plant matter. No one was breathing in enough oxygen and breathing out the carbon dioxide that plants needed for more photosynthesis. The oxygen level soared from about 15% of the atmosphere before the Carboniferous to 36%, a level at which even wet vegetation materials will burn. (The level is now 21%.) With no well-established terrestrial food chain, huge masses of unconsumed carbonaceous matter were piling up in peat bogs. In time, shallow seas covered them and overlaid them with mineral deposits. The peat was compacted into coal. This removed billions of tons of carbon from the biosphere and the climate turned cool. High temperatures at the beginning of the Carboniferous plummeted to as low as they would be until the ice ages.

Another source of global cooling was the weathering effect of plants on silicate rocks. Like thermophilitic bacteria before them, plants dissolve minerals as they break down rocks into soil, releasing carbon dioxide-absorbing chemicals into the air and water. This silicate weathering probably removed more carbon from the biosphere than photosynthesis and did more to cool the Carboniferous climate than the formation of fossil fuels.

The Earth's climate is sensitive to the carbon cycle, and life on Earth to the climate. Changes in atmospheric carbon that cause climate change are the result of relatively short-term disturbances of carbon moving through the biosphere (in and out of oceans, soil, biomass, and the air) and longer term disturbances of carbon moving in the lithosphere (in and out of fossil fuels and of carbonate rocks). Biospheric carbon levels tend to be self-regulating (high carbon levels encourage increased plant growth, which lowers carbon levels, etc.). Lithospheric disturbances are generally balanced, but not self-regulating. Large amounts of carbon dioxide are brought from the Earth's interior to the atmosphere by volcanic out-gassing, and large amounts are removed by silicate weathering, but there is no known feedback loop where one leads to another. It is a matter of chance when one or the other may happen. During the Carboniferous, the climate was dangerously cooled by disturbances in both the biospheric and lithospheric carbon cycles. In the biosphere, there were too many plants and not enough animals, so carbon was siphoned out of the biosphere in the form of coal. In the lithosphere, plant growth was causing silicates to tie up carbon in mineral form, and volcanoes were not bringing new carbon to the atmosphere. Temperatures fell, ice caps formed, and sea levels plummeted as the ice expanded. Plants grew and grew, packing more and more carbon into peat bogs and releasing more oxygen into the air. The Yin was way out in front of the Yang. But toward the end of the Carboniferous and beginning of the Permian, the Yang self-generated from within the Yin, and restored the balance. Before the icehouse effect seriously threatened the biosphere, overabundance of oxygen slowed plant growth and spurred forest and peat fires, releasing carbon dioxide back to the air.

The abundance of oxygen also encouraged animal life. The first vertebrate herbivores evolved, in the form of *Dicynodont*, a two-tusked mammal-like early reptile. It ranged in size from that of a small dog to a small rhinoceros, with tusks and powerful jaws. It spread throughout Pangea, feasting on the luxuriant growth of the supercontinent, breathing in oxygen and releasing carbon for the plants. New carnivores soon arose to feast on the herbivores, and a multi-tiered food chain and balanced ecosystem were on their way.

The new symbiosis was between plants and animals, but included bacteria and other microbes. Plant matter, particularly foliage, was high in carbon and low in protein. Plant eaters, both vertebrate and insect, depended on microbes to break down carbonaceous plant material in their digestive systems. As the microbes died, they contributed the protein in their own bodies to their hosts. Herbivores did not feed themselves directly; they fed microbes that fed them. Microbes seemed content to give themselves up to their symbiotic partners after a good life of digesting raw stems, leaves, and roots. The arrangement evolved first with insects feeding on decaying plant matter. The first reptile herbivores evolved from insectivores, picking up the microbial partnership along the way.

By late Permian times the Pangean supercontinent was fully formed and carbon dioxide levels had recovered sufficiently to bring a tropical, rainy climate to most coastal areas. Plants had plenty of water, soil, carbon, and sunshine. Sharks, shrimp, and trilobites filled the oceans, and insects filled the air. Microbes made grazers plentiful, and grazers made hunters plentiful. At least seventy-two species of reptiles roamed the land, some of them large herbivores like *dicynodon*, some small lizard-like insectivores, others, small carnivores. Twenty-five species of large, saber-toothed *gorgonopsians* were the carnivorous terrors of the late Permian. Three, four and five levels of predation wove the food web. There were no dinosaurs, mammals, birds, or flowering plants, and no bees, ants, or butterflies. But life on Earth was reaching a climax of species diversity. This was the Earth's first complex ecosystem.

We do not know why it ended, or how suddenly it ended. It may have happened overnight or over hundreds of thousands of years. The geological record shows a worldwide mass sedimentation at the end of the Permian, with a huge spike in fungal spores followed by a near disappearance of any sort of plant spores except ferns. There is also near universal oceanic anoxia (lack of oxygen), and a massive decline in animal species: terrestrial, marine, and aquatic. The sedimentation is evidence of a decimation of plant life followed by worldwide soil erosion. The fungal spore count is evidence of mushrooms feasting on dead plant matter followed by a proliferation of ferns and very little other plant life. Anoxia is evidence of shifting oceanic currents and overuse of oxygen by microbes in seawater caused by an overabundance of dead plant and animal tissue. The decline in animal fossils shows that whole families, orders, and classes of arthropods and other invertebrates died off, as did all large land animals, with the exception a genus of dicynodont, *lystrosaurus*, a moderate-sized herbivore that thrived and filled the empty world left by the extinction of its fellow reptiles. The earliest Triassic fossil beds following the event show 95 percent of the total animal population to be *lystrosaurus*. Eight orders of insects went extinct. Almost all sponges, crinoids, braciopods, foraminiferans, clams, snails, and fish died. But the fossil record is imprecise, however extensive. It is impossible to pinpoint events in much less than a million-year time frame. We do not know if the great P-T extinction was a single event or a series of catastrophes over a 600,000-year period.

We do know that the angel of death was global and thorough. It reached the deepest seas and shallowest inlets. It climbed up the longest rivers and the highest mountains, and passed over every great forest, swamp, and lake. Hills were stripped of topsoil down to bedrock. In the oceans, coral reefs died out, not to reappear for ten million years. Temperatures plummeted and then soared, acid rain fell everywhere, sea levels rose and fell. What could have caused all this?

The picture is sketchy at best, but the likely culprit is the Siberian Basalts Traps. These erupted exactly at the boundary between the Permian and Triassic periods, throwing up 2-4 million cubic kilometers of basalt lava. Lava flows 100 to 6000 meters thick covered an area the size of North America. These were not cone-forming volcanoes, but slow and steady upwellings of molten rock from the Earth's interior that flooded the landscape, releasing millions of tons of carbon dioxide, sulfur dioxide, fluorine, and chlorine into the atmosphere, all acid rain producers. Sulfur dioxide in the atmosphere forms sulfate aerosols that reflect sunlight, cooling the climate. The other gases released by the lava – flows, carbon dioxide, fluorine, and chlorine – are all greenhouse gasses, and fluorine and chlorine are killers in their own account. But the release of volcanic gases, though the largest in 600 million years, was not enough to cause the extinction on its own. Other factors were brought into the mix. The whole story probably begins with global *cooling* caused by the sulfur dioxide and by untold tons of sun-blocking volcanic ash and dust thrown into the upper atmosphere at the time of the eruption. This blanket of soot was enough to reduce or eliminate photosynthesis, while blocking the sun's warmth from reaching the earth.

Most of the Earth's vegetation was killed off, and species of plants unable to withstand a period of dormancy went extinct. Herbivores and carnivores would have run out of food in short order. The cooling lasted long enough for glaciers to form, tying up a large part of oceanic seawater. Sea levels fell, killing off shallow water species and sending massive amounts of decaying plant and animal matter to rob oxygen from deeper waters. Then the reverse happened. As the skies cleared of ash and sulfates, the greenhouse gasses kicked in. In addition to gases released by the lava flow itself, heat from the flow may have released untold quantities of carbon dioxide from the Tugusskaya coal beds that lay alongside the basalt. These coal beds were laid down during the Carboniferous and were the largest in world. Oxidation of this much coal would have been enough to accelerate the global warming process.

The climate turned hot, icecaps melted, and sea levels rose again. As temperatures soared, chlorine gas escaping from the lava flows may have drifted to the upper atmosphere and broken down the ozone layer, causing deadly infrared radiation to reach the Earth's surface. Then the real kicker hit. There was then, and is now, an enormous quantity of methane, another greenhouse gas, in the deep-sea polar regions of the Earth. Under extreme cold conditions and under intense pressure below 500 meters of ocean water, little cages of gas form, surrounded by water molecules. These are known as *methane hydrates*, and there are a lot of them. Recent estimates show them to contain more than twice the carbon of all fossil fuels on Earth. Either a lowering of sea levels reduced the pressures that contain the hydrates, or global warming melted enough of them to release an immense bubble of methane into the atmosphere. This would have caused a spike in overall greenhouse gasses in the atmosphere and led to a deadly acceleration of global warming. This extreme global warming, accompanied by a worldwide scourge of acid rain, would have killed the plants, the plant eaters, and the plant eater eaters, and stripped the land of topsoil. Dead plants and animals washing into the seas led to a further anoxia of deep ocean waters. The devastation of vegetation and the topsoil that sustains it was so complete that no new coal beds were laid down for 10 million years.

This model, or some variation, probably caused the great dying. It may have happened several times over an extended period. A little more of it and there would have been no recovery. As it was, the biosphere did not fully recover its species diversity until the age of dinosaurs, 50 to 100 million years later.

The K-T Event

The K-T extinction of 65 million years ago was only slightly less disastrous than the P-T. Where the P-T (Permian-Triassic) event 250 millions years ago marked the end of the Paleozoic ("Ancient life") era and the beginning of the Mesozoic ("Middle life") era, the K-T (Cretaceous-Tertiary) event marks the end of the Mesozoic and beginning of the Cenozoic ("Recent life") era. Each of the three eras (Paleo-, Meso-, and Cenozoic) is divided into several periods. The Mesozoic consists of three periods: Triassic, Jurassic, and Cretaceous. It is the K-T event 65 million years ago that killed the dinosaurs and made way for the current age of mammals.

Enough amphibians and reptiles survived the great extinction at the end of the Permian to radiate new species in the Triassic. *Lystrosaurus,* as we have seen, made it through and proliferated for a brief time throughout the supercontinent, but soon died out as new predators evolved. The first true mammals and the first primitive dinosaurs evolved in the Triassic, and the first birds in the Jurassic. As the age progressed, Pangea broke up into the separate continents we know today, creating isolated landmasses where evolution took separate courses. Of about a thousand dinosaur species, only a few dozen existed at any one time and place, as each species usually occupied a single landmass and endured an average of 5 to 10 million years before dying out. Dinosaurs originated in the Triassic and dominated the land, sea, and air in the Jurassic period. They climaxed in the Cretaceous. Mammals evolved at the same time, but remained small, mostly nocturnal, and scarce. For mammals this was a long, 150 million-year "fuse period" before their "explosion" in the Tertiary.

The Mesozoic was relatively stable in geology and climate. Temperatures and atmospheric carbon levels were higher than they are now and new coal beds were laid down. There were no ice caps, and sea levels rose to considerably higher than they are now. There were large, shallow seas covering inland areas of the continents, but these were not as extensive as they had been before the formation of the supercontinent. Colliding continents in the age of Pangea had created higher elevations and more relief, which, as the continents re-separated, diminished the area of inland seas. New mountain ranges – the Appalachians, Urals, and Atlas – had formed after North America pressed into Africa and Europe. The Rockies, Andes, Alps, and Himalayas had not as yet formed, and would not rise until the Tertiary collisions of the Americas with the Pacific plate, Africa with Europe, and India with Asia. The drifting apart of the continents throughout the Mesozoic allowed more moderate coastal temperatures to reach the central portions of landmasses and created oceanic currents between continents to mix tropical and polar waters, further moderating the global climate. Most dinosaurs occupied tropical and temperate regions, which were more extensive than today, but there were some species in polar regions. These probably migrated to temperate zones during the dark winter months.

The Jurassic was the age of the plated stegosaurs, armored ankylosaurs, and the *Iguanodon*, one of the first dinosaurs to be discovered (1822). Sauropods (some previously known as brontosaurs) grew 25 meters long and weighed 55 tons and more, the size of ten bull elephants. Ichthyosaurs were the reptile whales of the Jurassic, measuring up to 23 meters in length. These air-breathing monsters took to the seas 200 million years ago, after their amphibian ancestors emerged from the seas 200 million years earlier. Their numbers and size indicate a complex food chain in the oceans based on a healthy primary production of photosynthesizing marine plankton. Reptiles took to the air in the late Triassic in the form of pterosaurs, the first and largest ever flying vertebrates. In the Jurassic a reptile assumed the form of *Archaeopteryx*, and became the first bird. In the late Cretaceous *Tyrannosaurus*, the greatest predator of all time, hunted duck-billed hadrosaurs, horned ceratopians, and *Triceratops*. The very end of the Cretaceous saw the evolution of *Mosasourus,* a large swimming predator with a long neck and head above water. Its late appearance demonstrates that dinosaur orders were generating new species up to the very end.

The late Cretaceous was also the beginning of a new ecosystem based on birds, insects, mammals, and flowering plants. Plants had the great reproductive disadvantage of immobility: The apple does not fall far from the tree. To ensure cross-fertilization and dispersal, it was made to fall farther. It was so important for plants to spread their genes that they developed complex and ingenious strategies to do so that radically altered themselves and the rest of the biosphere in the process. Before the Cretaceous period plants depended on wind to scatter pollen and spores. Wind pollination worked best when plants were close enough to other members of their species to ensure fertilization. Distance radically lowers fertilization levels. A small percentage of pollen reaches its target in any case, and the wider spread the species, the lower the percentage. Once pollination was completed, a new round of wind dispersal took place. The spores that carried new plant life in this round of dispersal had to be light enough to be carried aloft. They could take very little along in the way of nutrition. Young plants that grew from spores were on their own from day one. They had to produce their own nutrition immediately. This was not good enough. Plants wanted to give their offspring a little something to get started. They needed to spread pollen without wasting it and needed to move spores far enough to ensure wide dispersal of the species. Animal mobility was the answer. Instead of developing legs of their own, plants hired the job out to the animal kingdom by making flowers. To spread pollen they used flowers to attract insects and later humming birds and bats. The key was to make the new bio-partners pick up pollen and spread it only to the *same species*. By growing flowers suitable to only a single species of insect, or flowering parts that only a particular species of humming bird could reach, they ensured that the courier had to be highly specialized as well, and would not waste its time, and their pollen, visiting the wrong species. This, in turn, led to a flourish of species diversity among insects and other pollinators. To pay for pollination services, plants produced enough pollen to feed their insect couriers in kind. Pollen, however, contains genetic material and is relatively "expensive" to produce, so some plants developed a secondary currency in nectar. Nectar is more amenable to both parties of the contract: sweeter to the taste of the pollinator and cheaper for plants to produce. Bees evolved along with other flying insects to fill this new market niche.

Seed dispersal was the job of birds and of the few mammals scurrying about among the dinosaurs. By equipping a plant embryo with a small food supply, plants not only gave their offspring a better chance of survival, they provided it with push. Most seeds were eaten along the way, but the few that survived more than made up for the loss. The trick was to allow animals to eat as much as they wanted without getting it all. One strategy was to produce enough seed to ensure that some is scattered without being eaten; another is to produce a protective coating around the seed that will get it through the animal's digestive system. This has the additional advantage of providing fertility for the young seedling. Another tactic is to provide something in addition to the seed for animals to eat, namely fruit. Fruit and nuts were a much more concentrated and protein rich food source than stems and foliage, and led to the radiation of hundreds of new warm-blooded animal species, particularly mammals, in the Tertiary period. The evolution of flowering plants also made it possible for both plants and animals to intersperse to a much greater extent than ever before. Even today grasslands and coniferous forests that depend on wind pollination are much lower in species diversity than flowering hardwood forests or tropical forests, where trees of the same species interbreed through insect pollinators and do not have to be next to one another.

The Cretaceous ended very abruptly. One day 65 million years ago the birds were singing, the bees humming, the dinosaurs roaring, and the next day it was all gone. An asteroid 10 to 15 kilometers in diameter slammed at 25,000 miles per hour into the Gulf of Mexico. An explosion millions of times the size of a hydrogen bomb shook the Earth. A tsunami over 1000 feet high rolled over Texas and the Gulf Coast. A hundred thousand cubic kilometers of dirt and debris were thrown into the atmosphere darkening the skies for months. Photosynthesis stopped, plants died, the earth went cold, and before it was over every large animal on the planet was dead.

About half of all living species became extinct, including all dinosaurs and most mammals. Rocks thrown by the impact into sub-orbit crashed back to Earth, igniting forests on every continent, burning half the planet's vegetation. Sudden heating of the atmosphere by the meteor's impact caused nitrogen and oxygen to combine into nitrous oxide and nitric acid, which fell as the worst acid rain the world had ever seen. The oceans were particularly hard hit. Ninety per cent of plankton species died out, eliminating many species higher on the food chain. Only two species of the Foramnifera phylum (an animal plankton) survived. Calcareous shells dissolved in acid. Coral reefs died out completely. Miraculously, many fish species survived. And then, as the skies cleared and the sun shone again, the climate turned hot. The impact caused enormous quantities of water vapor and carbon dioxide, both greenhouse gasses, to be released into the atmosphere.

Luckily for mammals, insects were virtually unaffected by the meteorite impact. Most insect species survived, which meant that insect eaters survived, and many insect eaters were mammals. Having cleared the Earth of large reptiles, the K-T event set the stage for a major new radiation of mammal orders in the Cenozoic Era.

The evidence is still fresh and we know much more about the K-T than the P-T event. There is good evidence that what I have described actually happened. But, interestingly, it might not have been enough. The picture is complicated by the fact that for several hundred thousand years around the time the meteor hit, the Deccan Basalt Traps were erupting in India, spewing the same sorts of climate busting gasses into the air as the Siberian Traps at the time of the P-T event. Though smaller than the Siberian Traps, they may have delivered the one-two punch. The picture is further complicated by evidence of a radical and unexplained *lowering* of sea level at the time.

THE MODERN EVENT

Recovery from the K-T event was fairly rapid. Ferns, always the first to colonize after a wildfire or volcano, covered the land where the forests had been as flowering plants slowly returned and ecosystems recovered. A new age of the great broadleaf forests began. The few species of sea plankton that survived the impact multiplied quickly and evolved into new species, re-establishing the oceanic food chain. Amphibians and surviving reptile families including turtles, lizards, snakes, and crocodiles repopulated the land. But the greatest radiation of new orders and new species was of mammals and birds. Only 3 families of mammals survived the K-T, but within a few million years there were 40, including several thousand new species. Whole new orders came into being, including rodents, bats, ungulates, and primates. In the early Tertiary, atmospheric carbon levels were high, the climate was warm, and there were no ice caps. Tropical forests stretched as far north as Canada and England. As temperatures moderated and rainfall decreased in mid continental regions, grasslands developed spawning a whole new array of grazing herbivores: horses, rhinos, pigs, hippos, cattle, deer, giraffes, camels, and antelope. This in turn bred new carnivores: dogs, bears, raccoons, and cats. As the climate cooled, ice caps formed at the poles and glaciers began creeping down mountain ranges. The world we know now was taking shape.

Deciduous forests, where trees and shrubs lose their leaves every year, originated not in the temperate zones where they are now, but in Polar Regions about 50 million years ago. The climate was warm and frost was virtually unknown, even in the far north and south, so there was no reason to drop leaves due to cold. Deciduousness, and the photo-period (day length) sensitivity that triggers it, evolved in response not to weather, but to the long winter darkness at the poles. It has since proven a useful adaptation to colder climates that destroy most types of plant foliage in winter.

The ice ages began around 2.5 million years ago and have been coming and going ever since. We are still in the midst of them, though we are currently in an "interglacial" period that is supposed to last another several thousand years. Possible causes of the ice ages are changes in solar radiation, a shifting of the Earth's orbit, or a re-routing of ocean currents. Ocean currents have changed as continents have drifted to new locations. There were no icecaps in the Mesozoic and early Cenozoic, probably because open oceans covered both Polar Regions and warm tropical waters circulated freely. But as Antarctica split from South America, Africa, Australia, and India, it drifted southward to its present position covering the South Pole. This blocked ocean currents from entering and mixing with the coldest polar waters, and also allowed snow to accumulate on the land. At the same time, North America and Eurasia drifted north and created a nearly land-locked Arctic Ocean that cut off most mixing with tropically warmed waters. Greenland, though not covering the pole, provided a platform for snow and ice accumulation. North and South America then joined at Panama, cutting off east-west currents. As ice began to build up over the years, it reached a positive feedback threshold: ice reflects sunlight, which creates more cooling, which creates more ice. Once the cycle started, it continued to feed itself until glaciers reached from the poles down to latitudes where tropical forests had been a few million years before.

A factor in creating the inter-glacial warmth we now enjoy is the formation, about 10,000 years ago, of the Gulf Stream. Cold, salty water from the North Atlantic is heavier than warmer, less salty water from the tropics, and sinks to great depths. It flows south along the bottom and then rises to the surface to be warmed again by the sun. This generates a reverse current northward of warm, lighter water along the surface, which keeps North America, and especially northern Europe, from freezing over. The current apparently started quite abruptly, and it could stop at any time. As the Greenland ice pack melts during the current period of global warming, fresh water running into the North Atlantic is not as dense and will not sink to the bottom, and may completely shut down the current. Without the Gulf Stream the North Polar region will become colder and the tropics hotter. Ironically, melting ice may trigger a new ice age.

Hominids and human beings have been around since the beginning of the Ice Ages, but avoided cold weather by remaining in Africa, southern Europe and southern Asia. *Homo sapiens,* the most recent hominid species, migrated northward into Europe and Asia as the most recent glaciers formed, and from Asia to Oceania and the Americas as the glaciers began to recede. As this newest of human species spread out across the land it developed the use of fire, new tools, and new hunting techniques. Sea levels were low and land bridges facilitated migration as ocean water was still tied up in ice masses. About 12,000 years ago the Clovis Culture, a technologically advanced people, made the land bridge crossing from Asia to the Americas, and populated most parts of both new world continents in about 2000 years. When these people arrived in Alaska the Americas held the most diverse array of animal species in the world. By the time they reached the tip of South America, most of the large animals were extinct. In North America the Clovis people killed all the wooly mammoths, all the mastodons and other elephants, all the giant bison, four species of camel, the saber-toothed cats, short-faced bear, giant armadillo, ground sloth, horses, and 28 genera of other large animals. Some of the large predators like the saber-toothed cats likely went extinct not by direct human predation but through human induced extinction of their prey. In South America the Clovis people killed every giant sloth and every giant armadillo, along with 44 other mammal genera. Less than a third of large mammal species were left. This began the mass extinction of the Modern Era.

What distinguishes the Clovis from earlier people is a six-inch long flint spear point. Near Naco, Arizona, a fossil mammoth was found with eight Clovis points embedded in its skeleton. With this tool and others like it, animals much bigger than people could be killed in relative safety. The Clovis people had other advantages, such as bows and arrows, clothing, language, communal hunting tactics, and domesticated dogs. But what they really had going for them in America was wildlife that had never seen anything like them. Animals did not know how to act around humans. In Africa, and to a lesser extent Europe and Asia, human cultures developed slowly as wild animal populations evolved around and along with them. Africa lost two genera at about the same time, and Europe seven, but most genera had time to adapt to human hunting, and survived. In the Americas, game species did not know what hit them.

That these animals went extinct 12,000 to 10,000 years ago is indisputable. We do not know for sure that humans were the only cause, but we do know that the extinction pattern happened a very short time after the arrival of humans. It affected only large, prey-sized animals. (Small mammals, reptiles, birds, and amphibians went untouched.) And America was not the only place it occurred. Wherever humans with advanced hunting technologies entered a region that had not known humans before, the same thing happened. In Australia, about 14,000 to 13,000 years ago the giant wombat, the giant kangaroo, and *Dromornis*, a three – meter tall flightless bird, all went extinct, as did 15 of 16 genera of large animals, a total of 45 species. One thousand years ago the Maoris reached New Zealand and exterminated the moa, another large bird, and people reaching the island of Madagascar at about the same time killed off all the lemurs and elephant birds. And we all know what happened to Auks and Dodo birds.

Stone Age hunting was the first of three waves of human-related extinctions. The second wave began 10,000 to 8000 years ago when people began clearing land for agriculture. Beginning in the Middle East and spreading to the temperate regions, vast expanses of grassland and hardwood forests fell to the ax and the plow, cleared of vegetation and animal habitat. The assault continues today in the tropics, where one percent of remaining rain forest is destroyed every year. The third wave began 250 years ago with Industrial Revolution and beginnings of fossil fuel combustion, climate change, and industrial pollution. Today somewhere in the neighborhood of 100 living species go extinct every day.

In the oceans small numbers of food fish remain. Whales are supposed to be protected, but are in danger of renewed hunting. Major oceanic kelp forests are dying out because there are too many sea urchins grazing on them. There are too many sea urchins because the large food fish that prey on them have been removed by over fishing. Coral reefs, the forests of the ocean that provide habitat for fish, crabs, shrimp, clams, sea fans, and sponges, began dying in 1960s and 70s. Estuarial habitats in coastal areas are degraded by sediment from agricultural runoff and urban development.

In the Appalachian Mountains of West Virginia and eastern Kentucky humanity has come up with a method of extracting coal known as *mountaintop removal*. Coal seams, usually at three or four levels in a single mountain, are only four or five feet high, and compose a small portion of the mountain's total mass. It is most efficient, from a financial standpoint, to extract this coal by removing the "overburden," or whatever lies around or on top of it. To do this, portions of the mountain are blasted with dynamite, the overburden bulldozed into an adjoining valley, and the coal trucked away. It takes about a year to remove a mountain. The process begins with the most species-diverse deciduous forest in North America, complete with oaks, pines, hickories, maples, beech, deer, turkey, raccoon, bear, bees, field mice, wasps, earthworms, ticks, chiggers, grasshoppers, trout lilies, bluebells, dandelions, poison ivy, jack-in-the-pulpits, wild iris, blue birds, summer tanagers, crows, hawks, eagles, cowbirds, and topsoil, or, in other words, overburden. Between one mountain and another runs a stream, with a living community of its own, of minnows, crawdads, frogs, salamanders and water bugs. This is where the overburden is dumped. Crushed rock filling the valleys leeches silt, sulfates, iron, copper, mercury, cadmium, and other heavy metals downstream. Mountaintop removal ends with a "reclamation" project that consists of piles of crushed rock planted with legumes and a few trees where patches of soil can be found. The coal is burned to generate electricity and its carbon released, restoring the hot early Carboniferous climate from which it originated. The biosphere is reduced.

In the past, tropical forests expanded or contracted in response to periods of alternating global warming and cooling. Periods of warming have been gradual enough for tropical forests to expand to higher altitudes and higher latitudes. Warming in the past has allowed forests to cover more of the Earth's surface, increase diversity and increase photosynthetic productivity. In the current warming,

however, tropical forests are not expanding, but contracting, with burning, logging, and clearing for agriculture and
development. Tropical species are going extinct before they
have a chance to migrate.

How bad is this? We do not know. We do not know how
many species are going extinct, how fast they are going, or
how many there are to begin with. We can't come close to
a good number. The number of discovered, named, and
described species stands somewhere between 1.5 and 1.8
million. Of these, a million are insects. Non-insect arthropods and other invertebrate animals make up around
175,000 more. Plants, most of them flowering plants, are
about 300,000. There are about 41,000 vertebrates (fish,
amphibians, reptiles, birds, and mammals), and the rest
are fungi, single-celled protists (protozoa), bacteria, and
viruses. But this is just what we know about. Every time
an entomologist shakes a tree in Amazonia another hundred new beetles species fall to the ground. Estimates of
the total number of species range from 5 million to over
100 million, and everywhere in between. We do not know
what is dying because we do not know what is living.

The P-T extinction was caused, as best we can tell, by
an Earth-based geological event, the eruption of the Siberian Basalt Traps. The K-T extinction was caused by an
extraterrestrial event, the impact of an asteroid. The cause
of the modern extinction is neither extraterrestrial nor geological, but biological. Its culprit is a single animal species
overwhelming other species and degrading the geophysical

foundation on which all species depend. It is life doing something to itself.

The current dying may turn out to be a "minor" mass extinction, similar in magnitude to dozens of others over the course of geological history. Or it may be a full-scale mass extinction, ranking right up there with the "big five." It may surpass even the P-T, sending life back to before the Cambrian explosion. There could be extinctions of whole classes and phyla, and perhaps of the animal kingdom as a whole, leading to an entire new beginning of multicellular life. Life has always had time to try new things, and the last 545 million years may turn out to be a trial run that will end the way the Ediacaran experiment ended just before it, leaving little more than fossil evidence of an evolutionary dead end. Perhaps the current extinction is the beginning of the end of life altogether. Who would be there to know that it had ever happened?

These are all firm possibilities. There is no discounting any of them, and no point in believing any of them either, as there is no way now to know which is closest to actuality. But there are some things that can be known and reasonable projections that can be made about what is going on now. In the remainder of this chapter I will make four observations and propose three alternative scenarios for the future.

First, the observations:

1) The biosphere is in crisis, the crisis is historically unique, and humanity is at its center.

Life adapts to change, but not to the rate of change we are seeing. Life will not continue as it is. Climate disruption, extinctions, and habitat destruction are happening so fast that life as it is will be sustained for at most a century or two. The crisis is unique in that life has never before gambled so much on a single species, never placed so much on a single roll of the dice. It has allowed one form to dominate all others and to disrupt the physical basis of all others. Before, life has depended on ecological balance, creating communities of interacting species that ebb and flow against one another, each keeping others in check. Single classes or orders have ruled the Earth, but never single species. Trilobites ruled the earth in the Paleozoic era, and dinosaurs in the Mesozoic, but there were thousands of kinds of trilobites and dinosaurs. None of them changed sea levels or altered the climate.

Humanity is at the center of the crisis, but it is short sighted, I believe, to say that humanity is the cause. We are a form of life driven to provide for itself as all life is driven to provide for itself: what we do, we do because we are alive. Humans eat, breathe, prosper, multiply, and defend themselves like every other form of life. This is what seems natural to most people. But life has put us in the forefront of all life – something we have not yet come to understand. We are not an external force acting against life; we are victims as much as other species. There is no separate *us* and *nature*. We do not yet have the metaphysical tools to understand the on-coming world, and can do little to effectively change what is happening. Self-blame is, in any case, a poor starting point for creative response.

2) Humanity is not ultimately responsible for life.

We are not as big as we often think ourselves. Humanity, though at the center of the current biological turmoil, is not yet critical to the survival of life as a whole. The extinction that began 10,000 years ago and continues today is not yet of end-Permian, or end-Cretaceous proportions. If humanity were to disappear from the scene now the biosphere would recover rapidly. Even if human civilization as it is were to persevere another two hundred years, with pollution, overpopulation, habitat destruction, soil erosion, resource depletion, etc., many species would eventually go extinct (including humans), but life would recover after a million years or so, perhaps in some dramatic new way. Severe and prolonged climate disruption would prolong the recovery period to tens of millions of years and perhaps bring the extinction to P-T proportions.

Far worse would be a nuclear war or series of nuclear wars at some time in the future. This is a firm possibility. Given political disunity of the human species, it is, in fact, a near certainty. Debris thrown into the upper atmosphere from a major exchange will block sunlight from reaching the Earth as it did during the eruption of the Siberian Basalt Traps and after the end-Cretaceous meteor, but radioactive fallout will do far more than either to destroy conditions for terrestrial life. A large enough war at some time in the future will generate radiation that will kill all higher plants and animals, and perhaps hamper DNA replication, even among single-celled organisms. Still, life would come back, somehow, after a hundred million or a billion years. Even if it did not – even if it were wiped out entirely – it would likely regenerate from scratch. Life has taken less than 4 billion years to get this far, and there are as many as 5 billion left before the sun runs out of fuel. The damage we do at our worst will be limited.

3) Human political disunity is a biological problem.

Humanity is less important to humans than faction or nationality. People everywhere, even in advanced societies, remain willing to kill other people, or to die trying, in the name of security for their own kind. Even in an age of intercontinental weapons and global communication, safety is thought to depend on keeping outsiders out, whether or not any sort of "out" remains available. Most people conceive of no other possible security arrangement. Fear of outsiders is so deeply engrained in the paradigm of civilization that it is considered endemic to any form of human organization. If this is true, there will continue to be warfare throughout the remainder of human existence.

As warfare becomes nuclear in the near future, it will become of biological as well as political interest. Persistent or prolonged exposure to radiation will have profound and permanent effects on the evolutionary potential of the biosphere as a whole. Developing a means of preventing war becomes, thereby, an adaptation for the living world as a whole. As war depends on political disunity, it can be prevented permanently only by political unity. Fear of outsiders will persist only as long as they continue to exist.

The long-term future of multicellular life depends on the near-term evolution of human political unity or the near-term extinction of human life.

4) Human understanding is a biological response.

Humans are the agents of change, but there are forces at work in the biosphere far below the surface of human thought and volition. Being human myself, I have no way to know what those forces are or what their evolutionary intent may be; I can only glimpse at what they are doing.

Part of what these forces are doing, I believe, is to make us understand. Understanding is itself a life force that is pushing and bulging its way to the surface amidst the wreckage of extinction and industrial pollution. Something new is arising – a sense of order that we have not seen or thought of before. It is not strictly scientific – perhaps not physical at all – though it is fueled and tempered by what we know through science. There is an entirely new emotional, almost visceral, concern for nature that transcends biological process. It is something that produces and reproduces itself, that reveals itself in the process of being perceived, and that creates itself from complexity and perplexity. Life's response to the mess it has made is *understanding*, and we are the designated understanders. Understanding is not usually understood in biological terms, but I think that is where it belongs. New ways of putting together what we see in nature, and in ourselves, is evolution itself. Appreciation of life is its hope. Life will become what it understands itself to be.

The following are three possible scenarios for the next hundred years or so: *stasis, catastrophe,* or *dominion:*

SCENARIO ONE, STASIS

Enlightened political leadership in major nations leads to legislation eliminating the worst forms of industrial pollution. Most countries limit their own population growth and develop cap-and-trade or carbon taxes to keep carbon emissions to a minimum. Consumers demand less, economic growth slows, alternative energy sources are developed, and fossil fuels consumption drops to a fraction of its former level. International treaties control nuclear weapons production and peace is maintained through a precarious balance of international power. Numbers of nuclear warheads are reduced, though emergency stockpiles remain in larger nations. Geopolitical tensions rise and fall with the tide of current events. Habitat destruction and species extinctions continue at a reduced rate, while new parks and nature preserves are established throughout the world. Humanity assumes a smaller role in the biosphere by taking up less living space and reducing its overall impact on the natural world. Natural cycles return in some areas and the climate begins to stabilize. Space exploration continues at a much slower pace, as few reasons are found to justify the expense. With periodic adjustments, life goes on at an even keel more or less indefinitely.

While the most attractive scenario, I consider stasis the least likely. It has the advantage of preserving fundamental concepts of who we are and promises continuity with the past and present, but it will not happen, I think, because it is not what life does. Life does not do stasis well – not without counterbalancing forces. It does not like to stand still or go back to where it has been, even if the past was a better place. Now that we are part of the carbon cycle, I do not think we can get out. Even if we use less fossil fuel and reforest large areas of the planet, we will have to monitor carbon dioxide levels indefinitely into the future and occasionally respond to imbalances before they get out of hand. People are addicted to continuous economic growth and do not feel organically whole without some form of it. Even with a radical alteration in what we mean by growth, we will continue to have growth. Nor do I believe it possible for humanity to remain politically divided for more than a few decades into the future. The international balance of power that has kept us from general war since the nuclear age began may continue to function for another generation or two but will not do so indefinitely. The balance will come unbalanced any number of times in the future, as it has in the past. Disarmament, as it is commonly understood, cannot be a permanent solution to the nuclear problem. Technology cannot be un-invented and we cannot go back to where we were before 1945. Nuclear weapons will be eliminated only when the political divisions that require them are eliminated. The symptoms will disappear only when the disease disappears. We will wonder why we ever had them.

Stasis is an unstated goal of the present day peace and environmental movements. Its perceived value is as a double negative: It is not pollution, not climate change, not over-population, and not nuclear war. But it fails to understand the onward momentum of civilization and of life itself. Stasis depends on a continuing coincidence of enlightened leadership in separate parts of the world and on an unrealistic inter-coordination of disparate national governments. If stasis were ever achieved, it would not last.

Scenario Two, Catastrophe

Despite attempts to regulate economic activities, environmental degradation continues. Progressive nations limit damage within their borders while others prioritize economic growth at the expense of industrial pollution, carbon emissions, and habitat destruction. Developing countries blame deteriorating global conditions on the earlier industrial development of wealthy nations, and claim the right to strive for economic parity. Carbon dioxide levels climb to new heights, sea levels rise, and global temperatures increase. Ocean currents become less reliable. Industrial production is concentrated in countries with poor labor laws and lax environmental protection. Populations stabilize in wealthier societies but increase steadily in underdeveloped areas, with immigration soaring as a result. Warfare remains the ultimate sacrifice of individual to society, and national governments resort to organized violence for the resolution of conflict. Public attention is distracted from global concerns by consumerism, fear, and battlefield heroics. A die off begins with plagues, wars, pollution, famine, and aggravated natural disasters.

Catastrophe, as I have described it here, is the most likely scenario because it is where we are headed. A straight line from where we have been through where we are now and on into the future points directly at resource depletion, environmental disaster, social upheaval, and nuclear war. Radioactive fallout will bring about the extinction of humans and most other large animals, along with many marine species and some advanced terrestrial plants. Insects, fungi, and plankton survival rates will depend on radiation levels.

Catastrophe is conceivable because it has happened before. Life does this sort of thing every hundred million years or so. The cause is unique but the effect is not new.

Scenario Three, Dominion

Stasis is humanity within the balance of nature.

Catastrophe is humanity out of balance with nature.

Dominion is the balance of nature balanced with the power of humanity.

Dominion means power. Humans have power over life on Earth. We dominate. We did not ask for dominion but it is given us. It is our condition. The continuing evolution of life depends on how we exercise it.

At this point we do not exercise it well. The world would be better if we did not have this power. Dominion has come to mean our right to slaughter animals and decimate forests without considerations beyond our own. It has come to mean the right to kill and pollute, to overpower the living world and bring it to the precipice of mass extinction. It has become the opposite of creation care. Yet we cannot shed it. It is the reality of life on Earth, and now, the hope for life on Earth.

There were many things people did not have to worry about until now: chemical balance in the oceans, rates of topsoil formation and erosion, ozone and carbon levels in the atmosphere, or ice melt in polar regions. We did not have to provide a place for the Spotted Owl and the Snail Darter. These took care of themselves; they were part of nature. If they did not take care of themselves, there was nothing we could do about it, or should do about it. But now we are moving into a new position. The future depends on conscious and deliberate management of natural processes. We have to know what we are doing and should be doing in relation to them. It is largely a matter of controlling ourselves – of stopping what we have been doing – but it is much more. We cannot go back to where we were. There will be no return to pre-nuclear, pre-digital, pre-industrial times. The force of life is not linear; it does not go backwards and forwards in time. We will do new things to grow into the role we are in: stop emitting so much carbon, but actively remove carbon we have already emitted; stop killing whales and elephants, but also create places for them to live. This is what I mean by *dominion*.

Dominion, for people, is an acceptance of power – of a degree of control with which we may not yet be comfortable. But it is less a matter of outright control than of positive interaction, of encouraging the balance one way or another. It is a meddling in affairs we have avoided before and an acceptance of new responsibility. But because it requires a blurring of the line between humanity and nature it means a less idyllic, less innocent relation to nature. It means there will be no ecological utopia. The climate will warm despite our efforts, species will go extinct, and remaining wilderness areas will no longer be as thoroughly wild as we would like them to be. Natural areas will exist by human designation and human definition, surrounded by human habitat. The Earth will be transformed from an open range to a garden plot, weeds and all. Wilderness will be contained by civilization, and civilization contained by itself. Humanity will assume a larger share of formerly natural processes and develop within itself the confidence to oversee the welfare of life as a whole. Nature will become less natural as humanity becomes more humane.

Adaptation to an environment of environmental imbalance may at some time in the future involve a degree of *geo-engineering*, or deliberate change in the physical basis of life on Earth. If, for instance, the climate becomes intolerably chaotic, humanity may decide to extract large amounts of carbon from the atmosphere by growing forests in places where they have not grown before. As I have suggested elsewhere, one way would be to plant trees on large floating platforms anchored offshore. Carbon dioxide would be absorbed in wood fiber and made available for lumber, relieving land-based forest ecosystems from timber production. But the risks would be enormous: shading of ocean waters would decrease plankton populations, lower water temperatures, and alter ocean currents. Atmospheric temperature changes would affect weather patterns, and too much carbon extraction would trigger an ice age, etc. The power to undertake geo-engineering projects may not come with the wisdom to undertake them successfully.

Life's adaptation to the environmental condition of dominion will depend on evolution of human awareness. To preserve life, we, the powerful, must know what is out there, and understand how large living systems interact. We will have to study life in new ways, develop new forms of expression and understanding, and create within the general public a big-picture awareness of ourselves in relation to the natural world. Power will be offset by a compassionate appreciation for the less powerful. We will learn to care for others as we care for the less powerful of our own kind. Life will evolve to a higher form as civilization develops the political tools it needs to act universally in relation to nature.

Dominion is more likely than stasis only in that stasis is next to impossible. It is less likely than catastrophe in that it is unknown and untried. We don't know what it is or could be because it has not existed. The force of life works best with small probabilities and large numbers: here it will be working with a small probability and one biosphere. A single misstep and the whole thing will come tumbling down. Practical ideas have to be tried in the real world and there is only one real world in which to try them. Each big idea we come up with will be a first and only try, and first tries rarely succeed in the evolutionary record. The risk of losing everything leaves no room for error, which leaves no room for trial and error. The path to dominion may lead to catastrophe at any moment.

Dominion is also less likely than catastrophe because it is idealistic: idealistic not in the sense of an imagined perfection but in the sense of arising in a conceptual rather than perceptual realm of consciousness. Because there is only one world to work with, dominion must be conceived before it is done. But it will not be *seen* before it is done: It will be a set of imagined possibilities understood in the collective mind before they are allowed to take root in the physical world. We will have to know what we are doing before we do it; and at this point, we do not know what we are doing. We know how to manipulate non-living systems but we do not know how to interact consciously with large, multi-parameter living systems, especially those of which we are a part.

Because it is unlikely, dominion is not an optimistic scenario. It is, however, hopeful. Hope is a form of awareness, an injection of spirit into vision of the future that brings meaning to the present. Hope does not require a calculated likelihood of success, but it enhances the likelihood of success by focusing energy toward a possible outcome. It provides direction in the present that leads toward a positive outcome even as the outcome itself cannot be known.

Dominion is both idealistic and hopeful. The same may be said for stasis, but dominion is more likely than stasis because it is practical. It could happen. To become practical it will require a level of creation care that is not yet evident, but there are some signs that we are moving in that direction. There is an arising awareness of who we are and what we should be doing. The soul will have its Kingdom.

Vision greater than seeing is hope.

Power greater than self is care.

Dominion is care for creation.

As the membrane of human civilization engulfs the plants and animals of the Earth and spreads to the depth of the waters and the height of the air, we come to see ourselves balanced with an overall balance of nature. We embrace self-control. We control industrial production and resource use. We limit our numbers. An awareness of our place in the larger world allows place for life that is not human. Vast expanses of the Earth are left for others, turning the membrane inside out. We do not abdicate our power. We do not wish for a simpler time. We restore a balance that was not previously under our control, a balance that we did not have to worry about before. We go forward in confidence. We *do* it; it does not happen on its own. This is dominion.

The human form will evolve slowly, but the structure of human consciousness will evolve rapidly. People will hear and see more in cyberspace than in the perceptual space surrounding their bodies. The physical space-time of perceptual consciousness will become a limited province within a wider-ranging spiritual consciousness. The Earth will no longer be the limit of life. The biosphere will replicate itself in small enclosures that spread out and infect the solar system and beyond.

VIII

The Kingdom

STRUCTURE

The kingdom was not originally an evolutionary concept. It was what you could see at a glance anywhere in the natural world. There were plants and there were animals and that was it. From Aristotle to the twentieth century the classification of living things developed around this fundamental dichotomy. Even as thousands of new single-celled organisms appeared under the microscope, scientists found themselves shoehorning everything into one kingdom or the other. Alga were clearly plants (immobility, photosynthesis) and amebas clearly animals (mobility, no photosynthesis), but then there were species like euglena, which swam around like animals *and* photosynthesized like plants. In the 1960s the suggestion arose as to the possibility of a third kingdom – that of all unicellular organisms. But even this did hold for long, as fundamental distinctions arose between prokes and eukes, and then between plants and fungi. There were so many new discoveries in biochemistry and evolution that classification shifted from a descriptive anatomical basis to an evolutionary basis. There are now a total of five recognized kingdoms: two unicellular: prokaryotes and eukaryotes, and three multicellular: plants, fungi, and animals. But this is by no means final. Research based on DNA analysis has thrown all traditional ideas of classification into question. We know too much now to be constricted by former categories of thought. Yet categories live on: there remain undeniable distinctions between banana trees, streptococci, algae, fungi, and prairie dogs. Five structural divisions of the living world are still recognized and continue to be called kingdoms.

Prokaryotes evolved, as we have seen, from amino acids and polymer chains. Once enclosed in membranes they became the first living cells (bacteria) and the first kingdom. For two billion years they were the only form of life on Earth. Rigid protein shields around each cell protected the chemical order on the inside from dissolution in the primordial soup on the outside. But the membrane, inflexible as it was, could not cut off the outside world entirely. Nutrients and metabolic compounds had to be allowed in and wastes allowed out. The membrane had to engage in chemical traffic to the extent of knowing how to direct it. The conscious experience of prokes, as best we can tell from the human perspective, was, and is, limited to the chemical realm of perception. Perception was limited to what we experience as *taste*. *Doing* was limited to deciding which chemicals to allow through the membrane and which to exclude, once they were tasted. Prokaryotes continue to live by absorbing organic compounds in the environment around them. A few, the cyanobacteria, have learned to manufacture energy directly from sunlight.

Certain types of prokes were engulfed within larger, more flexible collective membranes to become Eukaryotes, the second kingdom. Formerly independent bacterial cells, some photosynthetic and some not, prokes found themselves better off operating as specialized organelles within larger and more complex communities. Mitochondria and chloroplasts, originally free-living prokaryotic cells, retain to this day separate membranes and reproductive capacities within the eukaryotic cell. The more flexible and maneuverable outer membrane of the eukaryotic cell was able to sense the exterior world, to re-shape itself, and to engulf food particles in ways that a prokaryotic cell could never do. A new world of perceptual consciousness arose with the evolution of the eukaryotic membrane: the tactile realm. With it a new structure of consciousness evolved that coordinated tactile with chemical information. The sense of touch allowed eukes to experience external stimuli as tactile objects before experiencing them chemically. The evolutionary advantage was in developing a better sense of the shape of things beyond the membrane before allowing them through.

The evolution of the Eukaryotic Kingdom is the evolution of a new realm of consciousness and a new relation of that realm to the existing realm. The new realm is tactile perception; the existing realm is chemical perception. The relation is *time*. The concept of kingdom is usually understood from a functional standpoint; we attempt to understand it here as a new structure of consciousness.

All multicellular forms of life are combinations of eukaryotic cells. Fungi, the third kingdom, evolved as strings of non-photosynthesizing eukes discovered they were better able to move water and nutrients collectively than singly. They depended on external sources of metabolic energy, and though physiologically multicellular, they developed no sensory organs capable of perceiving beyond the chemical and tactile realms. The structure of their perceptual experience, therefore, remained that of the single cell. Eukaryotic cells that engulfed cyanobacteria as chloroplasts became algae. Algae live as single cells or as simple cell communities without extensive physiological interaction. But they must remain in water. To survive on dry they had to develop ways to pass water and nutrients from cell to cell. Plants, the resulting multicellular organisms, became the fourth kingdom. These were algae communities that evolved specialized organs to bring water from root cells to leaf cells, and carbohydrates from leaf cells to root cells. Plants are the only way to combine photosynthesis with life on land. But plants, like fungi, did not develop the capacity to perceive other than chemical or tactile stimuli.

The third and fourth kingdoms, therefore, though anatomically and physiologically multicellular, retain the same fundamental structure of perceptual consciousness as single eukaryotic cells. They remain chemo-tactile. They have no specialized sensory organs and no centralized processing of sensory information. A plant may react to dry environmental conditions by closing stomata, but the stimulus that closes the stomata is perceived and processed by individual guard cells, and not by the organism as a whole. The plant does not decide this, and does not do it. Individual cells within the plant perceive and react to the stimulus.

Animals, the fifth kingdom, have evolved from communities of non-photosynthesizing eukaryotic cells striving for greater efficiency in the pursuit of food and defense. The key feature of the animal evolution became mobility: moving the entire cell community toward food or away from becoming food. Successful mobility requires sensory organs, and with them the evolution of new realms of perceptual consciousness to process and make sense of perceptual information. Sensory organs produce for the animal kingdom new possibilities in which decisions have to be made and things have to be done. The organism as a whole has to perceive, think, and act. Consciousness is not merely multicellular, but organic; the animal becomes a conscious being greater than its cellular parts.

The evolution of the Animal Kingdom is the evolution of three new realms of perceptual consciousness and a new structural relation among them. The new realms are the olfactory, the auditory, and the visual. The coordination of these three is through *space*; their relation to the chemotactile realms is through *time*. The structure of animal consciousness is *space-time*.

The animal phyla that evolved in the Cambrian Era, or the vertebrate classes that evolved in the Permian, were entirely new forms of animal life, but not new kingdoms. There was no new structure of consciousness. Eukaryotic cells were arranged into creative new body forms to meet new evolutionary demands, but these levels of evolution did not turn existing forms into components of a higher order of being. New animal species were not at that point transformed into components of a superorganism. Sponges, worms, and tree frogs each evolved separate types of bodies – and thus separate phyla – but remained within a single kingdom. Each is a separate way of stacking the same basic eukaryotic components.

Each new kingdom is more than a new arrangement of existing components. It is a new level of organization in which existing forms of life – often the most highly evolved forms – *become* components. A new kingdom is a new dimension of organic wholeness over and above anything that existed before. Where the evolution of new species or families – even of classes and phyla – would be a new way to stack bricks into bigger, better, and more efficient buildings, the evolution of a new kingdom would be an arrangement of buildings into a city. Species, family, class, and phylum evolution involves a higher level of biological function; kingdom evolution involves a higher structure of consciousness.

It is unlikely that the Earth will ever see the evolution of new animal phyla. The thirty or so that remain are all that is left of the hundred that evolved during the Cambrian Era. Any successful new stacking of eukaryotic cells today would take millions of years of trial and error, during which time existing eukaryotic arrangements would be sure to devour them. We will have to be satisfied with the few basic body forms that are left – there is simply no room for new phylum level ideas. New species will evolve – perhaps new genera and families, but no new phyla and probably no new classes or orders either.

Where, then, will the force of life go? It cannot remain static in the face of the current ecological turmoil, yet it can come up with nothing fundamentally new within the animal kingdom. What can it come up with? A new kingdom of Eukaryotes? This is more unlikely than the evolution of new animal phyla, for the same reason. There will be too much competition from existing forms. What the force of life may do is to come up with an entirely new kingdom using, as it has before, existing forms as components. This is likely to be in the form of superorganism, an evolutionary direction already apparent in a few higher species of the Arthropod and Chordate phyla.

A new kingdom, to live up to its name, would have to incorporate existing species into an entirely new level of organization. Emerging ecological conditions would have to force a current life form into a more highly evolved arrangement with itself.

REALITY

That is how it will look from the outside. From the inside we see it coming already.

We think of tools as external to ourselves: they help us do things, but they are not us. But now, electronic tools – computers, telephones, and televisions – are changing what we see as well as what we do. They change consciousness. They feed us abstract ideas, information about far away places, and the internal workings of society. They convert perceptual to observational and other forms of consciousness. They change what we do, but more importantly, they change what we are.

Human consciousness is evolving with extreme rapidity. Most of what I know about the world I see and hear through electronic media. I see it on the screen. I see less with my own eyes directly and more through the eyes, cameras, and software programming of other people. More of what I know and think is known and thought by others. As more of my experience is electronically mediated, more is common to others of my species. What I saw on television last night was seen by millions of others, through the same camera angle. The experience was collective. This has happened in a hundred years or so – much of it in the last fifty or twenty years. The Cambrian Explosion took 5 to 10 million years to play itself out. We watch what is happening around us now, and marvel. Life itself is taking place on the screen.

I have used the video screen, in this book and others, as an analogy for real world perceptual consciousness. Real physical objects are like ranges of points on a multi-dimensional pixel screen, each point being a quantum location in several perceptual potentials. Objects we see around us are like icons – if you "click" where you see one of them, you can touch it or taste it there also. The analogy is good because it demonstrates how we perceive reality without material substance. We don't need anything external to consciousness for a sense of the real. We experience objects on television and computer screens knowing all the while that they are "no more than" patterns of activated pixels. We see a picture of something on the screen and say "That is a so and so!" knowing that it is "really" no more than appropriately arranged points of color. Real objects are similar arrangements in space-time. The *experience* of objects on a video screen or in space-time is true and certain, with no dependence on existence outside of consciousness. The real physical world is like a quantum pixel screen because, like information on the screen, it is nothing other than conscious experience.

I am turning that analogy around now by saying that consciousness is not only like the screen; the screen *is* consciousness. What you see on the television or computer is not just a picture; it is real – as real as the room around you. It is a new, evolving form of collective consciousness related to what is happening in the larger world of biology, the environment, and human history. It is a realm separate from perceptual consciousness but of the same primordial essence. It is the larger consciousness that we are becoming.

Objects and events on the pixel screen are not physically real. They may simulate the physical, but they have no mass, and their contours are not felt in the same dimensional shape in which they are seen. Virtual and physical realities are distinguishable by their structural context. The pixel screen is two dimensional, not three; its picture is more coarsely grained than the quanta of physical reality; it occupies only a portion of the visual realm; and there is no coordination between the visual and other perceptual realms, with the exception of the auditory. Screen reality is flat, its pixels are detectable, physical space surrounds its borders, and there are no touch, taste, or smell sensations associated with it. It is also *reducible* to the physical presence of the electronic interface. Screen reality *comes from* physical reality – it is there because there is a computer, a keyboard, and a plug in the wall. Experience on the pixel screen depends on its physical base and disappears when you pull the plug. It is as lifelike as life itself, but remains a small and relatively unimportant corner of overall perceptual experience.

This, of course, is what is changing so rapidly. We are becoming immersed in the screen, allowing ourselves be taken in and transformed by it. Television became a mass medium in the 1950s. People began watching, slowly at first, then for hours and hours every day. Community and national life began happening on the screen: Lee Harvey Oswald was assassinated in our living rooms and the Viet Nam War was on three channels every night. Nixon announced the Cambodian invasion on the evening news – within minutes I saw people pouring into the streets. Then, in the 1980s, home computers arrived. A friend of mine who owned a building supply store asked if I thought he would ever need one. In the 1990s the internet settled in among us – now nobody can pull himself away. Now we are texting and tweeting when we are supposed to be driving. Now we know what our friends are doing all over the world but don't know who the people are next door. With a cell phone you can talk to anybody in the world and see what they are seeing. You don't have to go anywhere anymore; all you have to do is know someone there. You see on the screen what they see with their eyes.

The screen is in the room, visible on the desk, or in the hand. There is a distance, still, between virtual and real reality. But life-sized holograms are on the way, and if you look under *Virtual Reality*, you will see 3D visors, virtual reality helmets, and data gloves for sale. There are even full-body data suits. You put them on and physical reality all but disappears. The screen is no longer flat; you step into it and you are in another world. The screen becomes the room. At this point the screen is fun and games, but soon it will be everything. How better to learn, to connect,

to know what it is like to be in someone else's shoes? How
better to see inside a building before you build it? How
better to network, to conference, to form a think tank, to
transcend the cares of the physical world? How better to
do your job without distraction? The physical interface is
still there and the distinction between the virtual and the
physical remains, but the shadow reality of one and the real
reality of the other are melting away fast. Each will become
a separate realm of human experience. Familiarity with the
virtual is the reality of the virtual. Its reduction to the physical interface seems increasingly less important.

Total immersion in artificial space is the new way to create the total interactive game experience. It may soon become the total problem solving experience as well, complete
with the gaming dynamic. Interactive virtual reality cells
may prove the way to study resource distribution, climate
change, and alternative economies. The social dynamics of
health service, employment, and housing may be more easily handled in an interactive gaming milieu. Having fun in
virtual reality may become the way to create life-fulfilling
social dynamics in the real world. Allowing large numbers
of people to play on line may become the way to test hy-
pothetical ecological, climate, and economic scenarios. It
may become the best way to present complicated bio-social
ideas, including those I have presented in this book. Whole
ganglia of interactive virtual reality cells, complete with
real-world feedback loops, may become interconnected at
higher levels to produce complex social policies that are

impossible to conceptualize in physical reality. New forms of virtual representation, or, better still, new forms of direct user participation may revolutionize the workings of democratic government. The current model of representative democracy – biannual bodily transport of voters to physical polling locations for the election of surrogates to rule for them – was designed for the horse and buggy world of the eighteenth century. The same system is still with us. Today there is no technical reason why anyone cannot immerse himself directly in interactive problems – social games – that particularly interest him, and participate in decisions. As electronic interaction is put to creative and serious use, the challenge will be to keep the system democratic by preserving equality of represented interests

But how will life on the screen relate to life in the room? To integrate the two worlds, there will be a more complex structure of consciousness. Physical space-time has always correlated to the tactile realm through the second time dimension. (You feel the g force as your body speeds up, goes over a bump, or around a curve in proportion to the magnitude of the second *per second* of acceleration.) But the relation of the same tactile realm to virtual space may be quite different, perhaps activating only portions of the surface of the body. Seeing and hearing in virtual space may be the same as in physical space, but *feeling* in most instances may be experienced in a less simulated manner. The chemical and olfactory realms may be coordinated at some point with virtual space. Most virtual space worlds are designed to simulate physical space to some extent, but *they do not have to*. Newer worlds may intentionally avoid space-time simulation, coming up with new forms of space – and time – adapted to new purposes. Any number of directions is possible. But each, whether it simulates physical reality or not, will coordinate with space in a manner that maintains the wholeness of consciousness. There will have to be some structural relation between the screen and the room. The information potential of each new virtual realm will have to coordinate with existing perceptual potentials, perhaps in the form of a new dimension or dimensions.

If this is the case, new realms of virtual reality will develop outside of the observational realm of consciousness. The observational realm is defined by physical and not virtual space-time: The observational information in words or numbers has to coordinate dimensionally with direct perceptual experience. You have to see what I say I see, and see it at the same place. Scientific progress, the systematic growth of the observational realm, depends on this strict adherence to dimensional structure. It will tolerate no deviation of space, time, and mass as they are coordinated into physical reality. Virtual reality, as it strays from physical space-time, will evolve outside of scientifically verifiable experience. It will remain a form of reality – a new realm or realms of consciousness – but coordinate with the observational realm in some as yet unknown manner.

This is to be expected in the evolution of a new kingdom. As new realms of consciousness become new realites, the information of which it consists remains reducible to pre-existing realms. The eukaryotic kingdom, for example, evolved when a new type of membrane enclosed existing prokaryotic components. As best we can tell, prokes were capable only of chemical perception. The new eukaryotic membrane, far more flexible and sensitive than the older prokaryotic membrane, created higher levels of internal metabolic order by evaluating external objects before allowing them to be engulfed. The cell perceived objects touching its exterior without engaging them chemically. This was achieved by monitoring the flow of potassium ions across the membrane's outer surface in the vicinity of the objects in contact with it. But because the potassium flow was itself chemical, a whole new world of perceptual experience evolved that was separate from, but reducible to, the chemical realm of perception. The information was chemical but the experience was tactile. In a similar manner, as eukaryotic communities evolved into animals at a much later evolutionary stage, the chemo-tactile experience of individual cells became the olfactory, auditory, and visual perception of the organism as a whole. Photons, for instance, were reducible to the tactile experience of retinal cells even as the visual realm took on a reality of its own. In each case the new kingdom meant new realms of consciousness reducible to existing realms. More recently, as the observational realm developed among human communities into an entirely new world of experience over and above what any individual can perceive directly, the information of which it consisted remained reducible to perceptual experience. Information in the form of sound symbols and ink patterns was still heard and seen, but the world it created was a reality many times larger than anything heard or seen directly.

The observational realm and virtual reality are both forms of what might be called collective consciousness. What is the relation, then, between them? Virtual reality is a different realm of consciousness, with a different kind of space. There may at some time be virtual reality systems that coordinate perfectly with all five perceptual realms, but the point is that they will not have to. They could as easily develop their own spatial structures that have nothing to do with non-visual perceptual realms. A virtual reality simulation of physical reality will remain just that – a simulation. Virtual space may look like physical space but it will not be the same thing. New realms of collective consciousness will evolve new forms of virtual space, but they will evolve alongside, and not within, physical space.

The physical interface – the visor, the data suit, the keyboard, the computer – will remain a structural distinction between virtual and physical reality. The virtual will remain reducible to, and dependent on, the physical – we will always be able to pull the plug. Even if a form of virtual reality becomes something we cannot live without – even if it becomes the world we live in – it will remain dependent on physical reality. Our bodies will have to stay warm, fed, reproduced, and healthy. This parallels the dependence of multi-cellular on unicellular consciousness. Our cells have to stay warm, fed, reproduced, and healthy in order to maintain consciousness on an organic level. If they don't get what they need, they pull the plug.

Familiarity with virtual space will mean less reliance on physical space as an absolute structure of an external universe. As new forms of space arise, physical space will seem less unique, and consciousness will no longer seem contained by it. The outer edges of physical reality will appear against the deep background of a larger metaphysical context. Space-time will be a limited sphere – a special structure of consciousness that evolved to coordinate the perceptual experience of complex cell communities during the Cambrian Period. The new realms of cyberspace opening now to humanity parallel in many ways the new realms of perceptual experience that opened to eukaryotic cells as they became animals.

This is the reality of our time. It is unavoidable but not inevitable. It may fail. It may begin to take shape around us, and then not happen. The force of life is with us, but may not stay with us.

Soul

The Kingdom is not just evolution and technology.

The Kingdom may prove a wondrous new age or a devouring monolith. We may wish it or fear it, but we will not decide for or against it. It is happening to us. The motion is of being as a whole – good, bad, and neither, a motion far larger than humanity, far larger than anything we can do or decide to do. Cyanobacteria did not decide to be engulfed by eukarotes, nor flagellates to be regimented into sponges. The force of life is doing its thing, and there is no way to start or stop it. Decisions will be in how to adapt.

The river will not be pushed or pulled; we can only guide our craft in the stream.

To be one with the stream – to be what we are becoming – something will have to happen to the human spirit. We will have to understand being in a new way. We will see consciousness beyond the self, becoming a living form that has not yet existed. A spark will ignite the human spirit.

It will be magic. There is no logic in the self transcending itself.

The Kingdom will fail without its soul because evolution and technology will create only stronger and more highly integrated versions of what we already are. We will extract more, produce more, and pollute more. We will procreate more, exploit more, degrade more, and leave less for other forms of life. We will go where we are now headed. We will build smarter means of destroying one another, and we will use them. We will be too powerful to not destroy ourselves; and we will think, because of the way we think, that it is the right thing to do. There will be failure of the current chapter of life on Earth.

Yet the force of life is with us. It is behind the power we have amassed. It has given into our care its highest organic accomplishments and millions of years of its evolutionary creativity. Large animals will survive if we care for them to survive. Forests and oceans will live if we see the life in them and see that it is good. The weather will give life if we come to see that it has life to give. It is we who sentence life or death. The force of life depends on us to do what it is trying to do.

But we cannot do it as we are.

Look at what life is. Listen to the sounds. Feel the joints and muscles in your body. Feel the blood in your veins. Watch the breath passing in and out. Thoughts flicker through the mind – think them; then just watch them. Watch thoughts without thinking them. Look past what you think to what you do not think about. Look at what is there, whatever it is. Look at the rocks, the trees, and the soil, the weeds and the bugs. Hear the birds. Look at the little things you feel. Whether or not you need them, look at the little things. Do not do anything – just be. Look at everything there is.

We think we are human because we think. Thought is what you share with people. You do not think what they think, but you are one of them because you think. You share breath with plants and heartbeat with animals. You are they, in body and in soul. You feel; they feel. Your bones are the Earth, and your blood, the oceans. They are you. The part of you that is human is thought.

The future of humanity matters little to life as a whole. Millions of years will be wasted, but there are hundreds of millions more. The Earth will heal. Life will find new direction.

To identify with life is to diminish the self-importance of humanity.

To know life beyond humanity is to create a place among the living for humans.

Now think. Think of what we must do to live. Think of climate chaos, vanishing forests, dwindling resources, soil washing down gullies, missiles waiting in silos, millions born without food or clean water. To do, we have to think, to become fully human. It is a special part of what we are.

We do not yet know how to think. We do not yet identify as humans. We are divided against ourselves. We have no soul as humans. Everything there is to do must be done with everyone, yet there is no wholeness. Each of us has more important things to be than human.

Despair comes of lacking the means to do. If the environment, nuclear weapons, and overpopulation were items on the list that we had not yet gotten around to, the adrenaline would soon surge in our veins and we would go to work. We would sense the challenge, the danger, the excitement, and heroes would rise in our midst. But we have no tools to begin work. Humanity is an empty identity. We are two hundred separate nations fearful of outsiders. The atmosphere has no jurisdiction and the oceans lie beyond anyone's responsibility. Industry flows to countries that do not protect labor, pollution flows to countries that do not protect the environment. Population policy is a violation of somebody's nationally sovereignty – a conspiracy of outsiders. Technology gravitates toward new ways of killing people, yet nobody is thinking about real-world, down-to-earth conflict resolution. Nobody dares think outside the nationalist paradigm. Weapons of total destruction spring up everywhere and we do not know how to prevent war. We will watch it happen and wonder why. Humanity wants to prevent war, but *we* do not want to prevent war; it is too important to who we think we are.

There is no basic approach to any of the crises facing us, yet each of them is enough to end human civilization. Despair fills the human spirit because there is no way to act. There is no way to begin to act. The power given us is pushing us to destruction because we do no know who we are.

The Kingdom will not come without its soul.

It belongs to no formula, no ideology, no faction, no set of opinions that must be believed. It is beyond the confines of logic or science, unpulled by reason or gravity. It is beyond all understanding, yet known to everyone. It is hope. It is why there is life.

The Kingdom is the unseen justice of trunk, root, and leaf,

The love of one for the other:

Doing as you would have it done to you.

It is the sacred hoop, the flowering tree: a dream enacted, unseen, in the light of day.

It is the multitude, amazed at the foot of the mountain, believing, entering.

It is two or three gathered together.

We will be quarrelsome as ever, but together. We will be divided together: divided without violence, for violence is not viable to what we have become. The force of life, when we see it, will force us from the violence of creed and nationality, from the separateness of ourselves; that we might live.

The Kingdom is the spirit of organic function.

Soul is the illogic of non-self.

The Kingdom is in our mines and mills, in the hum of our highways, in the stream of pipes and wires beneath our city streets.

It is the chemistry of commerce and

The biology of industrial production.

It is the organic imperative of justice.

The spring has begun to fade. Trout lilies have blossomed by the stream and melted into the foliage. The bunting and grosbeak are back. I sit on the porch, resting from my work as life again rises from the ground. A soft rain waters seeds sprouting in the garden. Layers of dark leaves mold and crumble in the soil, offering atoms of last year's sunshine. Carbon feeds nitrogen; billions live through their days and die; amino acids course through the earth to my body, and back.

The children are grown and gone now, planting gardens of their own in currency and toil. They are no longer with us, and I do not see them in my day. They are our magic. I do not know how it is that they came into this world.

The owl sings, "Who cooks for you?" by the riverbank. The force of life is a great doing, a self: a great curvature of the soul.

BIBLIOGRAPHY

Cells

Knoll, Andrew H., *Life on a Young Planet*, Princeton University Press, Princeton NJ, 2003.

Southwood, T.R.E., *The Story of Life*, Oxford University Press, 2003.

Phyla

Berquist, Patrick R., *Sponges*, University of California Press, Berkeley, 1978.

Brusca, Richard C. and Gary J., Invertebrates, second edition, Sinauer Associates, Publisher, Sunderland MA 2003.

Buchsbaum, Ralph, *Animals Without Backbones: An Introduction to the Invertebrates* University of Chicago Press, 1976.

Holldobler, Bert, and Wilson, Edward O., *Journey to the Ants: A Story of Scientific Exploration*, The Belnap Press of Harvard University, Cambridge, MA, 1994.

Hoyt, Erich, *The Earth Dwellers: Adventures in the Land of Ants*, Simon & Schuster, New York, 1996.

Maeterlink, M. (1905), *The Life of the Bee*, (English trans. By A. Sutro, Dodd, Mead, and Co., New York, p. 427.

Margulis, Lynn, and Schwartz, Karlene V., *Five Kingdoms*: An Illustrated Guide to the Phyla on Earth, W.H. Freeman and Company, New York, 1988.

Schneirla, T.C., *Army Ants: A Study in Social Organization*, W.H. Freeman and Company, San Francisco, 1971.

Sudd, John H., and Franks, Nigel R., *The Behavioural Ecology of Ants*, Blackie, Glasgow and London, 1987.

Valentine, James W., *On the Origin of Phyla*, The University of Chicago Press, 2004.

Wheeler, W. M. "The Ant Colony as an Organism," Journal of Morphology, Vol 22, no 2, pp 307-325, 1911.

Wilson, Edward O., *The Insect Societies*, The Belnap Press of Harvard University, Cambridge, MA, 1971.

PLUMBING

Blake, Marion Elizabeth, *Ancient Roman Construction in Italy*, Carnegie Institution of Wahington, publication 570, 1947.

Evans, Harry B., *Water Distribution in Ancient Rome*, The University of Michigan Press, Ann Arbor, 1994.

Granger, Frank, *Vitruvius on Architecture*, William Heinemann Ltd., London, G.P. Putnam's Sons, New York, 1934.

Herschel, Clemens, *The Two Books on the Water Supply of the City of Rome of Sextus Julius Frontinus, Water commissioner of the City of Rome A.D. 97,* Dana Estes and Company, Boston, 1899.

Landels, J.G. *Engineering in the Ancient World*, Chatto and Windus, 1978.

FORTIFICATION

DeVries, Kelly, *Medieval Military Technology*, Broadway Press, Peterborough, Ontario, 2003.

Hogg, Ian, V., Fortress: *A History of Military Defence*, St. Martins Press, New York, 1975.

Kenyon, John R. *Medieval Fortifications*, St. Martins Press, New York, 1990.

Otterbein, Keith F., *How War Began*, Texas A & M University Press, College Station, 2004.

Toy, Sidney, *A History of Fortification from 3000 B.C. t0 A.D. 1700*, The MacMillan Company, New York, 1955.

Tracy, James. D. *City Walls: The Urban Enceinte in Global Perspective*, Cambridge University Press, New York, 2000.

TELEGRAPHY

Holtzmann, Gerard J., *Early History of Data Newworks*, IEEE Computer Society Press, Los Alamitos, California, 1995.

Taylor, William B., *An Historical Sketch of Henry's Contribution to the Electro-Magnetic Telegraph*, p 54. in Shiers, *The Electric Telegraph, an historical Anthology*.

Silverman, Kenneth, *Lightning Man: The Accursed Life of Samuel F.B. Morse*, Alfred A. Knopf, New York, 2003.

REPRODUCTION

Adelman, Saul J., and Adelman, Benjamin, *Bond for the Stars*, Prentice Hall, Englewood Cliffs, NJ, 1981.

Appelbaum, Robert, and Sweet, John Wood, eds, *Envisioning an English Empire*, University of Pennsylvania Press, Phildelphia, 2005.

Bridenbaugh, Carl, *Jamestown 1544-1699*, Oxford University Press, New York, 1980.

Harrison, Albert A., *Spacefaring*, University of California Press, Berkeley, 2001. TL 1500.H37 2001.

Heppenheimer, T. A., Colonies in Space, Stackpole Books, Harrisburg PA, 1977.

Macvey, John W., *Colonizing Other Worlds*, Stein and Day, Briarcliff Manor, NY, 1984.

Tyler, Lyon Gardiner, *Narratives of Early Virginia 1606 – 1625*, Charles Scribner's Sons, New York, 1930.

Zubrin, Robert, *Entering Space,* Jeremy P. Tarcher/Putnam, New York, 1999.

DEATH ON EARTH

Benton, Michael J. *When Life Nearly Died*, Thames and Hudson, London, 2003. QE 721.2 E97 B45 2003.

Boulton, Michael, *Extinction: Evolution and the End of Man*, Columbia University Press, New York, 2002. QE 721.2 E97 B68 2002.

Briggs, Derek E.G., and Crowther, Peter R., *Paleobiology II*, Blackwell Science, Oxford, 2001.

Erwin, Douglas H., *Extinction*: How Life on Earth Nearly Ended 250 Million Years Ago, Princeton University Press, Princeton, 2006.

Gould, Stephen Jay, *Wonderful Life: The Burgess Shale and the Nature of History*, W. W. Norton and Company, New York, 1989.

Hallam, Tony, *Catastrophes and Lesser Calamities*: The Causes of Mass Extinction, Oxford University Press, New York, 2004.

Hallam, A., and Wignall, P.B., *Mass Extinctions and Their Aftermath*, Oxford University Press, New York, 1997.

McMenamin, Dianna L.S., and McMenamin, Mark A.S., *Hypersea*: Life on Land, Columbia University Press, New York, 1994.

McMenamin, Dianna L.S., and McMenamin, Mark A.S., *The Emergence of Animals*: The Cambrian Breakthrough, Columbia University Press, New York, 1990.

Southwood, T.R.E., *The Story of Life*, Oxford University Press, New York, 2003. QH 325 S688 2003.

Ward, Peter D., *Rivers in Time*, Columbia University Press, New York, 2000.

VIRTUAL REALITY

Germain, Gil, *Spirits in the Material World*: The challenge of Technology, Lexington Books, Plymouth UK, 2009.

Nusselder, Andre, *Interface Fantasy:* A Lacanian Cyborg Onthology, MIT Press, Cambridge MA, 2009.

Hein, Michael, *The Metaphysics of Virtual Reality*, Oxford University Press, New York, 1993.

Thompson, Richard L., *Maya: The World as Virtual Reality*, Govardhan Hill Publishing, Alachua, Florida, 2003.

Castronova, Edward, *Exodus to the Virtual World*, Palgrave Macmillan, New York, 2007.

Boellstorff, Tom, *Coming of Age in Second Life*, An Anthropologist Explores the Virtually Human, Princeton University Press, Princeton NJ, 2008.

ENDNOTES

1. The term *observational* is derived not from biology, but from physics. The idea of a dimensional structure of consciousness occurred to me upon consideration of the "role of the observer" in quantum mechanics and relativity theory.

2. This assumes five, not six, life kingdoms. Some taxonomies consider Archaeans to be a kingdom separate from Prokaryotes.

3. Knoll, p. 128-138.

4. Knoll, p. 206-224.

5. Bergquist, pp.74-78.

6. Bergquist, Patricia R., *Sponges*, University of California Press, Berkeley, 1978: diagram, p. 63. (QL 371.B47 1978b).

7. Buchsbaum, p.79.

8. The "highest" members of the three "highest" phyla show some interesting traits. Cephalapod mollusks have opted for advanced sensory organs and nervous development where hymenopteran arthropods have given this up in favor of social development. Primate chordates seem to be trying to combine the two.

9. There are thirty-four or so recognized animal phyla, the best known being *Porifera* (sponges), *Cnidaria* (hydras, jellyfish,

coral, sea anemones), *Platyhelminthes* (flatworms), *Nematoda* (roundworms), *Annelida* (segmented worms), *Mollusca* (clams, snails, squids, octopuses) *Echinodermata* (starfish, sea urchins), *arthropoda* (lobsters, crabs, insects, spiders), and *chordata* (vertebrates: fish, amphibians, retiles, birds, and mammals). An evolutionary trend from "lower" to "higher" phyla cannot be denied, but it should be pointed out that most evolutionary "progress" has occurred not by one phyla evolving into another, but among species within phyla. The first multicellular animals probably developed some time in the Precambrian or early Cambrian period (600 million years ago), and "quickly" (within about 100 million years) branched out into the basic body plans that we recognize as phyla. Our phylum (Chordata) is nearly as old as that of sponges and flatworms. They are our cousins, not our forebears.

10. Brusca, Richard C., and Brusca, Gary J. *Invertebrates*, Second Edition, Sinauer Associates, Inc., Publishers, Sunderland, MA, 2003, p462.

11. Holldobler, Friederike, and Wilson, Renee, Journey to the Ants: A Story of Scientific Exploration, The Belknap Press of Harvard University, Cambridge, MA, 1994, pp. 98-100.

12. Wilson, The Insect Societies, pp. 228-231.

13. Wheeler,W.M., "The Ant Colony as an Organism," Journal of Morphology, 22(2):307-325, (1911).

14. Maeterlinck, M., (1905), *The Life of the Bee*, (English trans. By A. Sutro), Dodd, Mead, and Co., New York, 427pp.

15. Wilson. P. 318.

16. The relation of the sensory realms to spatial and temporal dimensions is discussed in more detail in Avery, Samuel, *The Dimensional Structure of Consciousness* and *Transcendence of the Western Mind*.

17. There are three ordinary dimensions of space and two of time. The second time dimension corresponds to *mass*, which

is expressed in terms of acceleration, or meters per second *per second.*

18. This means that the space-time structure of light is adopted as a universal medium for all perceptual experience, not just for vision.

19. The idea of matter existing independent of perceptual experience is due to the perception of a physical object in more than one perceptual realm at the same location in space and time. But it is an assumption; it can be neither proven nor disproven. Scientists generally assume the existence of material substance, but it is unnecessary to their work.

20. Granger, p.143.

21. Herschel, P. 197. ("Nat Hist" xxxi. 24).

22. Evans, p.20.

23. Kenyon, p. 16.

24. Tracy, p. 26.

25. Toy, p. 182.

26. Toy, p. 143.

27. Toy, p. 238.

28. Holzmann, p. 72.

29. Holzmann, p. 89.

30. Holzmann, p. 174.

31. Taylor, William B., *An Historical Sketch of Henry's Contribution to the Electro-Magnetic Telegraph*, p.54, in Shiers, *The Electric Telegraph, an Historical Anthology.*

32. From the Record of the Supreme Court of the United States, taken at Boston, September, 1849, in the case of Morse vs. O'Reilly, in Shiers, *The Electric Telegraph, an Historical Anthology*, p. 113.

33. Silverman, pp. 240-242.

34. Bridenbaugh, p. 119.

35. Bridenbaugh. p. 136.

36. Captain Nathaniel Butler, in Tyler, p.413.

37. Tyler, p. 416.

38. Tyler, p. 322.

39. Appelbaum, p. 275.

40. Appelbaum, p. 44.

41. Appelbaum. p. 28.

42. Smith mentions this episode several years after the fact, but does not included it in the original description of his captivity under the Powhatans.

43. Tyler, p. 316.

44. From *Proceedings of the English Colony*, Tyler, p. 169-173.

45. There is no documented proof of this, of course. But Opecancanough was reported to be around 100 years old in 1644, which would make him 17 when Menendez took him to Spain. His acceptance into the ruling family upon his return in well documented, the brother then ruling was no doubt Wahunsunacock.

46. Bridenbaugh, pp 10-17.

47. Tyler, p.348.

48. Zubrin, p. 109.

49. In the midst of writing and re-writing this section I opened a fortune cookie at a restaurant that said: "If the brain were so simple we could understand it, we would be so simple we couldn't."

50. For a complete explanation of the correspondence between dimensions and perceptual realms, see Avery, Samuel, *The Dimensional Structure of Consciousness*, or Avery, Samuel, *The Quantum Screen*.

51. McMenamin, *Hypersea*, p. 224-226.

www.ingramcontent.com/pod-product-compliance
Lightning Source LLC
Chambersburg PA
CBHW062148080426
42734CB00010B/1612